Stable Recursions

With Applications to the Numerical Solution of Stiff Systems

COMPUTATIONAL MATHEMATICS AND APPLICATIONS

Series Editor

J. R. WHITEMAN

Institute of Computational Mathematics, Brunel University, England.

E. HINTON and D. R. J. OWEN: Finite Element Programming

M. A. JASWON and G. T. SYMM: Integral Equation Methods in Potential Theory and Elastostatics

J. R. CASH: Stable Recursions: with applications to the numerical solution of stiff systems

Stable Recursions

With Applications to the Numerical Solution of Stiff Systems

J. R. CASH

Department of Mathematics
Imperial College of Science and Technology, London

1979

ACADEMIC PRESS

A subsidiary of Harcourt Brace, Jovanovich, Publishers

London New York Toronto Sydney San Francisco

ACADEMIC PRESS INC. (LONDON) LTD.
24/28 Oval Road,
London NW1

United States Edition published by
ACADEMIC PRESS INC.
111 Fifth Avenue
New York, New York 10003

British Library Cataloguing in Publication Data

Cash, J R
Stable recursions.—(Computational mathematics and applications).
1. Differential equations—Numerical solutions
2. Recursive functions
I. Title II. Series
515'.352 QA372 79-50521
ISBN 0-12-163050-1

Printed in Gt Britain by Page Bros (Norwich) Ltd., Norwich

Preface

The demands of modern technology have produced a need for scientists to solve numerical initial and boundary value problems involving systems of ordinary differential equations. As a result a multitude of finite difference techniques have been proposed for dealing with these problems. The advent of modern computers gave many of these techniques wide application and consequently a set of *ad hoc* procedures which were known to be useful for practical computation, in some rather vague sense, were developed. For a long time, however, the theoretical aspects of these finite difference schemes remained neglected. Probably the first really successful attempt to establish finite difference schemes on a firm theoretical basis arose from the work of Dahlquist and Henrici. The latter author's book, published in 1962, has become a cornerstone for the basic theory of discretisation methods for ordinary differential equations. In a more recent book Stetter (1973) has produced a general framework which encompasses many of these methods and, as a consequence, the classical $h \to 0$ theory of discretisation methods seems to be in a relatively satisfactory state. However, both Henrici and Stetter, amongst many others, have noted that there is at least one class of equations, namely stiff systems of equations, for which this analysis is inappropriate and which requires rather special treatment.

A great deal of the early theory relating to stiff systems of equations was developed by Dahlquist. This culminated in his classic theorem that "An A-stable linear multistep method must be implicit and must have an order of accuracy less than or equal to 2". Implicit in the proof of Dahlquist's theorems, although not usually stated when his results are quoted, is the requirement that the desired solution of the linear multistep method in question has to be generated directly in the conventional way. It is usually the case that it is better to use an efficient direct method, if available, rather than an indirect one. However, as an immediate consequence of Dahlquist's and similar theorems, most reasonable finite difference methods for the numerical solution of stiff systems of non-linear ordinary differential equations are essentially iterative. Thus it is natural to ask whether or not improvements in the stability properties of our schemes could be obtained if the solution were to be generated from the outset using an iterative technique. As we shall show in Chapter 3, if the required solution is generated using a precisely defined iteration scheme acting on the linear multistep method itself, it is possible to derive schemes which are A-stable and which have an order of accuracy greater than 2 and also to derive convergent linear k-step methods with orders of accuracy greater than $2[k/2] + 2$.

In this text we examine various classes of iteration schemes in the hope of developing finite difference procedures which can be used for the efficient numerical integration of systems of ordinary differential equations. It is pointed out that it is not meaningful merely to state whether or not a particular linear multistep method is stable. The precise way in which the required solution is to be generated must be stated and only then can the stability properties of the complete method (= discretisation scheme + precise method of generating the required solution) be discussed.

The text is intended mainly for scientific research workers and requires very few pre-requisites apart from a knowledge of some of the basic analytic properties of difference and differential equations. It is not meant as an introduction to finite difference methods but it is hoped that the material presented will help the reader to deepen his understanding of this particular class of methods. One of the key observations is that the well known class of finite difference schemes based on linear multistep methods, with which the reader should be familiar, is merely a particular case of a much wider class of integration procedures which are examined in detail in the forthcoming chapters.

As the title suggests, my main aim in writing this book has been to bring together algorithms for the efficient numerical integration of stiff systems of ordinary differential equations since this is one particular area where the existing theory is far from satisfactory. At several points in the text, however, I have departed from this basic aim and have considered the development of algorithms for the solution of various non-stiff systems in cases where it has seemed that an iterative approach may be valuable. Several of the algorithms have proved difficult to analyse with complete mathematical rigour and it has been necessary to rely almost entirely on numerical evidence to demonstrate their potential. I have, however, not worried too much if an algorithm is not backed up by the necessary mathematical theory. Rather I have proposed the algorithm in the hope that this will stimulate further research. In view of this, the present text is not in a "cut and dried" form, but contains numerous open ended questions which seem to merit further investigation. Indeed, for this reason I have at several points deliberately suggested algorithms with virtually no theoretical backing (but with numerical results to demonstrate their potential usefulness). At many points in the text I have illustrated particular algorithms by applying them to fairly simple test problems, since it is often the case that the actual mechanics of an algorithm become more transparent if it is applied to a simple, rather than a complicated problem. I have, however, been careful to ensure that the algorithms derived also perform reasonably well on more complicated problems of practical interest.

In Chapter 1 some of the relevant theory relating to the stability properties of discretisation schemes and linear recurrence relations is introduced. Classically the theories of finite difference integration schemes and recurrence relations have been developed separately and one of the main purposes of this text is to examine and exploit as much as possible the interrelationships between these two theories. In Chapter 2 we examine in more detail the theory behind the numerical solution of linear recurrence relations, since many of our ideas for the solution of ordinary differential equations come from this source. In the first part of Chapter 2 we consider Olver's algorithm for the numerical solution of second order linear recurrence relations and examine its extension to vector equations. Although Olver's algorithm is extremely efficient for this particular class of problems it does seem to have the drawback that it cannot be extended directly to deal with non-linear equations. In order to overcome this difficulty an iterative method of solution is derived and this has the advantage that, as well as being reasonably efficient for linear problems, it may be extended directly to an important class of non-linear equations. In Section 2.5 we consider the extension of these iterative techniques to the solution of higher order recurrence relations and as a result a class of algorithms based on the Gauss-Seidel approach is constructed. In the non-linear case it is necessary to use two distinct iteration schemes, which we have referred to as the primary and the secondary schemes, and it is shown that a one-step Gauss-Seidel–modified-Newton scheme is efficient for the solution of our non-linear equations. Finally in Chapter 2 we consider a completely different class of iteration schemes which do not have as wide a range of application as those considered earlier but which do have very important practical applications in the numerical solution of ordinary differential equations.

In Chapter 3 we consider the extension of some of the recurrence relation techniques derived in the previous chapter to the numerical solution of systems of ordinary differential equations. Although, as previously mentioned, our main interest is in developing algorithms for the numerical solution of stiff systems, for the sake of completeness we start our analysis in Chapter 3 with a discussion of the non-stiff case. A class of algorithms based on one-step Gauss-Seidel–direct-iteration schemes is developed and it is shown that these are often satisfactory for the solution of our problem. Also considered in Section 3.2 is the class of conventional Adams predictor–corrector schemes used in PECE mode and it is shown that these schemes may be extended in a natural way to yield a more general class of Adams integration schemes. In Section 3.4 we examine in more detail the stability aspects of our iterative integration procedures. It will have become clear by the time that this stage of the text is reached that the existing theory is not sufficient to cover iterative methods, since different ways of generating the required solution often give rise to methods with differing stability properties. A new concept of iterative absolute stability is derived and this helps us to recognise iterative algorithms suitable for the integration of ordinary differential equations. In Section 3.5 a class of iterative algorithms based on a Gauss–Seidel approach is developed for the integration of stiff systems. Although we are not quite able to achieve full A-stability, a class of schemes which behave like either one- or two-step methods, have high orders of accuracy, and have infinite regions of absolute stability is derived and this class is shown to be useful for the integration of stiff systems. In Section 3.6 we consider a somewhat different approach based on the class of iteration procedures developed in Section 2.7. This approach is rather unusual in that the finite difference schemes in question are not solved exactly and the primary iteration scheme is only applied a fixed number of times—usually once. Also in this section a close correspondence is derived between our algorithms and the method of deferred correction originally developed by Fox. The approach does have practical significance since it allows us to develop A-stable schemes of order greater than two. In Section 3.12 we consider the extension of some of the techniques developed earlier in Chapter 3 to deal with the numerical solution of second order equations. The algorithms derived do not have a firm theoretical basis and as a result it is necessary to rely mainly on numerical evidence to demonstrate their potential. In particular an attempt is made to extend Olver's analysis to derive "quadratic factors" of high order linear recurrence relations and this is another area where further research is called for. Finally, in Section 3.13 some of the iterative techniques developed for ordinary differential equations are extended to deal with the heat conduction equation with non-derivative boundary conditions. Although only relatively few numerical experiments have been carried out with these particular algorithms, the results obtained do seem to indicate that the approach is promising. It is once more hoped that the algorithms will serve to stimulate further research, although there is an obvious need for additional theoretical results.

In Chapter 4 we consider a radically different approach to the solution of stiff systems. We extend some of the procedures developed in Chapter 2 to produce a class of iterative algorithms which are applied directly to the system of differential equations itself rather than to an approximating discretisation method. This class of procedures has a much more limited range of application than the algorithms derived in previous chapters, since it requires the system to have a special structure. For systems which do possess this structure, and these are quite common in practice, the algorithms are particularly powerful since they do not involve truncation error. In the second half of Chapter 4 we consider the relationships between these iterative algorithms and certain other widely used methods based on singular perturbations and pseudo-steady-state approximations. The main practical difficulties associated with these last two classes of methods are discussed in some detail and it is shown how the algorithms developed in the first half of Chapter 4 overcome most of these difficulties. Finally, in the last part of Chapter 4 we consider algorithms for the solution of certain classes of second order equations which cannot easily be solved using finite difference techniques. In particular an efficient algorithm is derived for the solution of certain problems having a Jacobian matrix possessing at least one eigenvalue with a large imaginary part.

I should like to take this opportunity of expressing my thanks to the numerous people who have influenced my thinking on this subject. I was fortunate indeed to be able to spend some time with Professor Hans Stetter in Vienna and without his constant help and encouragement during the early stages of the preparation of the manuscript this book would surely never have been written. Also I should like to mention the tremendous help received from Professor F. W. J. Olver and Drs R. V. M. Zahar and J. C. P. Miller, all of whom taught me a great deal about recurrence relations. I am pleased to be able to thank Professor Richard Bellman for asking me to write this book, Professor John Whiteman for placing this book in his series and for his tremendous help in editing the manuscript, Academic Press for their technical advice during the preparation of the manuscript and Mrs Sandra Place for her excellent typing from an almost unintelligible script. Finally I would like to thank my wife, Roslyn, for her love and understanding during the time spent writing this book and (if I may steal an apt sentiment from the introduction of a book by H. J. Stetter, 1973) if there is anyone at all who will rejoice at the appearance of this text it will be her.

January 1978. J.C.

Contents

1

Some Interrelationships Between the Theories of Recurrence Relations and Ordinary Differential Equations

1.1 Introduction

Traditionally the techniques and theories relating to the solution of linear recurrence relations and to the numerical solution of initial value problems in ordinary differential equations (O.D.E.s) using linear multistep methods, although often closely related, have usually developed along slightly different lines. Algorithms for the numerical solution of linear recurrence relations are usually designed so as to be able to generate *any* desired solution which is uniquely specified in advance by means of certain initial conditions (even though this problem may be very badly posed initially). However, when solving O.D.E.s using finite difference methods it is usual to assume that the required solution is in some sense *dominant*, at least for small values of the steplength of integration. The main purposes of the present text are to examine in some detail the interrelationship between these two theories and also to develop a different approach to the numerical solution of O.D.E.s by exploiting the theory of linear recurrence relations. In particular we shall develop classes of iterative techniques which do not demand that the required solution of the system of differential equations being solved is in any sense a dominant solution of the finite difference scheme being used and which converge rapidly to the required solution for a large set of initial approximations. These techniques produce satisfactory results for the integration of stiff initial value problems using linear multistep methods even when these are not zero-stable. This approach also enables us to develop schemes which are based on linear multistep methods and which are *A*-stable with order greater than two.

As an introduction this chapter starts with a brief survey of some of the relevant theory relating to the solution of linear recurrence relations and ends with an analysis of some of the stability aspects of linear multistep methods.

1.2 Some results from the theory of linear recurrence relations

Consider first the solution of the kth order linear recurrence relation

$$\sum_{j=0}^{k} a_j(n) y_{n+j} = g(n), \tag{1.1}$$

where the $a_j(n)$ and $g(n)$ are given sequences of the non-negative integer variable n with $a_k(n) \neq 0$ for any n in the region of interest. Any solution, y_n, of (1.1) may be expressed in the general form

$$y_n = \sum_{i=1}^{k} c_i y_{i,n} + p_n, \tag{1.2}$$

where the $y_{i,n}$, $i = 1(1)k$, are any k linearly independent solutions of the homogeneous part of (1.1)—i.e. equation (1.1) with $g(n) \equiv 0$—p_n is any particular solution of (1.1) and the c_i are arbitrary constants. The linearly independent set $(y_{1,n}, \ldots, y_{k,n})$ will henceforth be referred to as a *basis* for the solution of (1.1), and any member $y_{i,n}$ of this set will be called a basis solution. Any basis solution $y_{1,n}$ having the property that

$$\left.\begin{aligned} &\lim_{n\to\infty} (y_{i,n}/y_{1,n}) = 0 \text{ for all } i = 2, 3, \ldots, k \\ &\text{and} \\ &\lim_{n\to\infty} (p_n/y_{1,n}) = 0 \end{aligned}\right\} \tag{1.3}$$

will be called *dominant at infinity* or simply *dominant* and it will be assumed that it is possible to choose p_n and the $y_{i,n}$ so that there is a basis solution, $y_{1,n}$, which is dominant. Most of the analysis could be extended to include the case where there are two or more independent dominant solutions, e.g. where the largest roots of the characteristic equation (in the constant coefficient case) comprise a conjugate complex pair. We shall, however, restrict ourselves to the case where there is a single dominant solution since, as will be seen later, this is the case that is of most concern when differential equations are integrated using linear multistep methods. Any solution y_n of (1.1) containing a non-zero multiple of $y_{1,n}$ will be called a dominant solution. When this is not so, y_n will be called a non-dominant solution of (1.1). If any k consecutive values of $y_j, j = 0(1)k-1$, are given initially, the constants c_i in (1.2) are completely specified and a unique solution of (1.1) is defined. This unique solution could be generated recursively from the given initial conditions by using (1.1) in the form

$$y_{n+k} = (1/a_k(n))\left\{g(n) - \sum_{j=0}^{k-1} a_j(n)y_{n+j}\right\} \qquad n = 0,1,2,\ldots, \tag{1.4}$$

where the right hand side of (1.4) is known when needed for all n. Equation (1.4) will be referred to as a *direct method* for generating the solution of (1.1).

Although, as previously mentioned, the required solution of (1.1) is in theory completely specified by any given k consecutive values of y_n, an unstable build up of rounding errors can often render (1.4) completely ineffective as a numerical method for the solution of (1.1). This is invariably so if the required solution is a non-dominant one. This can be seen if it is assumed that we require to compute a solution $\bar{y}_n$ of (1.1) which is such that

$$\bar{y}_n = \sum_{i=2}^{k} c_i y_{i,n} + p_n, \qquad c_2 \neq 0.$$

The effect of rounding, which is in general inevitable on an automatic machine, means that a perturbed solution $\hat{y}_n$ of (1.1) is computed, and this has the form

$$\hat{y}_n = \sum_{i=1}^{k} \hat{c}_i y_{i,n} + p_n, \qquad \hat{c}_1 \neq 0.$$

In the early stages, the term $\hat{c}_1 y_{1,n}$ is very small and has a magnitude which depends on the accuracy to which the calculations are performed. The relative error, e_n, in $\bar{y}_n$ is given by

$$\begin{aligned} e_n &\equiv (\bar{y}_n - \hat{y}_n)/\bar{y}_n, \\ &= \left\{\sum_{i=1}^{k} (c_i - \hat{c}_i)y_{i,n}\right\} \Big/ \left\{\sum_{i=2}^{k} c_i y_{i,n} + p_n\right\}, \\ &= \left\{-\hat{c}_1 + \sum_{i=2}^{k} (c_i - \hat{c}_i)y_{i,n}/y_{1,n}\right\} \Big/ \left\{\sum_{i=2}^{k} c_i y_{i,n}/y_{1,n} + p_n/y_{1,n}\right\}. \end{aligned}$$

From (1.3) it follows that $|e_n| \to \infty$ as $n \to \infty$. From this we see that the relative error in our approximation to $\bar{y}_n$ increases without bound as $n \to \infty$. If $c_j, j > 1$, is the first non-zero coefficient appearing in (1.2) then the calculation, using (1.4), of any $\bar{y}_n$ approximating the y_n of (1.2) will be liable to such a large accumulation of round-off error as to render the whole procedure totally ineffective. Gautschi (1967) has given some spectacular examples of such instability. Clearly there is a need for special methods for approximating y_n if any solution other than a dominant one is required. This has motivated recent research into the development of algorithms for computing non-dominant solutions of (1.1), which has been mainly directed at replacing the original initial value problem by a mathematically equivalent, well conditioned boundary value problem. Such a replacement may be done by abandoning an appropriate number, $k - m$, of the initial conditions and requiring instead that the solution should vanish at $k - m$ points sufficiently far outside the region of interest. It is very important to note that we do not require the solution which we are computing necessarily to tend to zero for large n. (Indeed, it is often possible using this approach to compute non-dominant solutions which increase without bound as n increases.)

The requirement that the desired solution should vanish outside the range of interest, in the absence of any further information, is a convenient computational procedure. However, if we do have some knowledge of how the required solution behaves for large n, such as via an asymptotic expansion, it is usually better to use this information to define new boundary conditions rather than simply to set them equal to zero. In certain circumstances it can be shown that the solution of this re-posed boundary value problem is a good approximation to the required solution of the original initial value problem over any given finite range providing the new boundary conditions are set "sufficiently far" outside the far end of this range. The case $k = 2$ of linear second order recurrence relations will be considered in the next chapter. There it will be shown that, before any solution values have been computed, the optimal positioning of the boundary conditions to obtain a given degree of precision over a specified range for any particular problem can be determined. For the case $k > 2$, however, the optimal positioning of the boundary values *ab initio* is still largely an unsolved problem. In order to recognise that direct methods of solution are unstable, and to be able to select an appropriate value for m, it is necessary in general to have some additional knowledge regarding the behaviour of the required solution. This knowledge need not necessarily be enough to enable the boundary conditions to be set more accurately, but it should be enough to enable the wrong solution to be detected during a computation. As a result, the theory for the case $k > 2$ is mainly of use in the field of ordinary differential equations. An important practical case is that in which a basis may be chosen so that $y_{i,n}$ dominates

$y_{i+1,n}$ for $i = 1, 2, \ldots, k-1$ as n increases, in the special sense that

$$|y_{i,n+1}/y_{i,n}| > |y_{i+1,n+1}/y_{i+1,n}| \qquad (i = 1, 2, \ldots, k-1; n = 1, 2, \ldots),$$

where $|y_{i+1,n}/y_{i,n}| \to 0$ as $n \to \infty$ for all i.

In this case it is usually possible to state that the re-posed problem does have a solution closely approximating that of the original problem (Oliver, 1968a). Furthermore for this situation, all the complementary functions may be ranked in a complete hierarchy of dominance with the basis solution $y_{i,n}$ being the ith in the order. Equations for which we are able to pick out a dominant basis solution have important practical applications in the numerical solution of initial value problems for ordinary differential equations using linear multistep methods. The implications of this will be considered later. However, we first examine some of the analogies between the classical stability theory relating to linear multistep methods and certain parts of the basic theory of linear recurrence relations as outlined above.

1.3 Some stability aspects of linear multistep methods

We consider first of all the numerical solution, using a linear k-step method, of the system of first order ordinary differential equations

$$\frac{d\mathbf{y}}{dx} = \mathbf{f}(x, \mathbf{y}), \tag{1.5}$$

with initial condition $\mathbf{y}(x_0) = \mathbf{y}_0$. The linear k-step method has the form

$$\sum_{j=0}^{k} \alpha_j y_{n+j} = h \sum_{j=0}^{k} \beta_j f_{n+j}, \tag{1.6}$$

where α_j and β_j are constants and where it is assumed that $\alpha_k \neq 0$ and that α_0 and β_0 are not both zero. Here we have used a notation consistent with common usage so that h is a positive constant called the steplength of integration, $\mathbf{y}_{n+j}$ is the approximate numerical solution obtained at the point $x_{n+j} = x_0 + (n+j)h$ and $\mathbf{f}_{n+j} = \mathbf{f}(x_{n+j}, \mathbf{y}_{n+j})$. Equation (1.6) will also be normalised in the usual way by assuming that $\alpha_k = 1$. If we now introduce the first and second characteristic polynomials, denoted by $\rho(\delta)$ and $\sigma(\delta)$ respectively, associated with (1.6) as

$$\rho(\delta) = \sum_{j=0}^{k} \alpha_j \delta^j,$$

$$\sigma(\delta) = \sum_{j=0}^{k} \beta_j \delta^j,$$

it is well known that scheme (1.6) is consistent if and only if

$$\rho(1) = 0, \left.\frac{d\rho(\delta)}{d\delta}\right|_{\delta=1} = \sigma(1).$$

An immediate implication of this is that for a linear multistep method to be consistent the first characteristic polynomial $\rho(\delta)$ must have a root at $+1$. The use of the word roots, rather than zeros, in this context has been adopted to conform with current usage. It is usual to refer to this root as the *principal root* (denoted by δ_1). All other roots, which can only arise if $k > 1$, are called *spurious roots*, since they do not correspond to a solution of the original differential equation. In order that we may generate a solution of (1.6), it is usual to assume that k initial conditions of the form $\mathbf{y}(x_j) = \mathbf{y}_j$ are given, or more realistically have been computed using an alternative starting procedure if $k > 1$. Equation (1.6) may then be solved for $\mathbf{y}_{n+k}$ by rewriting it in the form

$$\mathbf{y}_{n+k} - h\beta_k\mathbf{f}_{n+k} = \sum_{j=0}^{k-1} (h\beta_j\mathbf{f}_{n+j} - \alpha_j\mathbf{y}_{n+j}), \tag{1.7}$$

where the right hand side of this expression is known. If the original system of differential equations is linear, or if (1.6) is explicit, equation (1.7) may be solved for $\mathbf{y}_{n+k}$ immediately. If (1.5) is non-linear and (1.6) is implicit, however, it will in general be necessary to solve for the required solution $\mathbf{y}_{n+k}$ using some sort of iterative technique. It is assumed that h is sufficiently small for the particular iteration scheme being used to converge to $\mathbf{y}_{n+k}$. In either case we shall refer to scheme (1.7) as being a *direct method* for the solution of (1.6), since $\mathbf{y}_{n+k}$ is determined exclusively in terms of the previous k known values of $\mathbf{y}_i$ and does not depend on values of $\mathbf{y}_i$ for $i > n + k$. By using scheme (1.7) for $n = 0, 1, 2, \ldots$ in succession, we may generate a solution of (1.6) at all steppoints in the range of interest. If we wish to generate a solution of (1.6) using a direct method, we have to be sure that the solution which we actually obtain is the required solution of (1.5) and not a spurious solution of (1.6)—i.e. corresponding to one or more spurious roots of $\rho(\delta)$—which bears no relation to the solution of (1.5). We note that in the case $h = 0$ we can always distinguish clearly between the principal root and the spurious roots of $\rho(\delta)$. However, as will be seen later, in the case $h > 0$ we are not always sure which root is the principal root and so it is necessary to proceed with caution. As we have already shown in our analysis of linear recurrence relations, there may in some cases be an unstable accumulation of errors arising as a result of perturbations from the required solution. This could, for example, be due to the effect of rounding and truncation errors. It is usual in the treatment of discretisation methods to refer to *perturbation insensitivity* as *stability*. Among the weakest of stability conditions which are usually imposed on (1.6) is that of *zero-stability* which essentially describes what would happen in the limit $h = 0$ and which may be defined as follows.

Definition 1.1

The linear multistep method (1.6) is said to be zero-stable if all roots of $\rho(\delta)$ have modulus less than or equal to unity with all roots of modulus unity being simple.

The restriction of zero-stability, which is sometimes referred to as *D-stability*, serves to ensure that the required solution of (1.6) (with $h = 0$) is not dominated by any of the spurious solutions. If this is not so, as we have seen previously, there will be an unstable build up of rounding error rendering the whole scheme totally ineffective over anything but a very short range of integration. One of the basic theorems relating the highest attainable order of a zero-stable linear multistep method to its stepnumber k is the following, the proof of which may be found in Henrici (1962).

Theorem 1.1

No zero-stable linear multistep method of stepnumber k can have order exceeding $k + 1$ when k is odd or exceeding $k + 2$ when k is even.

If we now consider the case $h \neq 0$, which is of course the only important case in practice, scheme (1.6) will in general be non-linear. Consequently it will not be possible to develop a theory relating to the order of dominance of solutions of (1.6). In order to be able to overcome this problem and to be able to obtain some useful results regarding the stability properties of (1.6) with $h \neq 0$, it is usual to apply (1.6) to the scalar equation

$$\frac{dy}{dx} = \lambda y. \tag{1.8}$$

In (1.8) λ is a complex constant with negative real part so that the equation is inherently stable for all initial conditions. In a practical case λ may be regarded as a piecewise constant approximation to $\partial f/\partial y$ if f is a scalar function, or to an eigenvalue of the Jacobian matrix $\partial \mathbf{f}/\partial \mathbf{y}$ if $\mathbf{f}$ is a vector-valued function. If we apply the linear multistep method (1.6) to the solution of (1.8), the resulting recurrence relation is again linear, has constant coefficients and has stepnumber k. In this more general case a stability condition usually imposed on (1.6) is that of *absolute stability* which may be defined as follows.

Definition 1.2
The linear multistep method (1.6) is said to be absolutely stable for a given value of $q = h\lambda$ if, for that value of q, all roots r_i of the stability polynomial

$$\rho(r) - q\sigma(r) = 0 \tag{1.9}$$

satisfy $|r_i| < 1$ for all $i = 1, 2, \ldots, k$.

This condition ensures that any spurious solutions of (1.6) which are brought into the solution of (1.6) which we actually compute, due to rounding error or truncation errors, decay with increasing n. As explained previously a spurious solution is one which does not correspond to the solution of the original differential equation. However, this decay condition does not always ensure that we obtain the required solution over a large range of x. An unacceptable build up of relative error may ultimately occur if the required solution decays faster than one of the spurious solutions. A closely related concept to absolute stability is that of *relative stability*, which serves to ensure that the spurious solutions decay faster than the required solution. As $q \to 0$ the roots of (1.9) tend to the zeros of the first characteristic polynomial $\rho(\delta)$. The conditions of consistency and zero-stability together imply that $\rho(\delta)$ has a simple zero, which is denoted by δ_1, at $+1$. We denote by r_1 the root of the polynomial (1.9) which tends to δ_1 as $q \to 0$. It is assumed that this root is well defined for all complex constants q with $\operatorname{Re}(q) < 0$. This assumption is far from trivial since there may be cases where, as q changes, two roots including the principal root of (1.9) first become identical and then part again. It will generally not be clear in these cases which root we need to follow as our principal root. A definition of *relative stability*, as in Hull and Newbery (1962), is now given.

Definition 1.3
The linear multistep method (1.6) is said to be relatively stable for a given value of q if, for that value of q, the roots r_i of the stability polynomial satisfy

$$|r_i| < |r_1|, \qquad i = 2, 3, \ldots, k.$$

Although they do not state the fact explicitly, Hull and Newbery assume that there is never any difficulty in picking out the principal root from the solutions of (1.9). The conditions of zero- and relative-stability merely demand that the required solution of our linear multistep method is not dominated by any of the unwanted spurious solutions. We find it necessary to impose stability restrictions of this nature since, if there is a solution dominating the required solution of equation (1.6), our direct method is not effective over a sufficiently long range of integration. However, we may remove the assumption that the required solution is to be generated using a direct method, and determine it instead using a boundary value technique. In this case it will no longer be necessary to impose the conditions of zero- and relative-stability and, as will be shown later, perfectly valid solutions may be obtained using non-zero stable schemes. One of the main purposes of the next two chapters will be to examine methods, other than direct ones, by which the required solution of (1.6) may be generated. It will be shown that it is possible to develop schemes which behave like linear k-step methods, in that they need only k initial conditions of the form $y(x_i) = y_i$, but which are convergent with order exceeding $k + 1$ when k is odd and exceeding $k + 2$ when k is even.

Stability problems become even more acute if the system (1.5) exhibits the property of *stiffness*. For the purposes of this section we shall define stiffness in the following way. (See Lambert (1973).)

Definition 1.4
The linear system of ordinary differential equations

$$\mathbf{y}' = \mathbf{A}\mathbf{y} + \mathbf{p}(x), \qquad \mathbf{y} \in \mathbb{R}^s$$

is said to be stiff if

$$\text{(i)} \quad \operatorname{Re}(\lambda_j) < 0, \qquad j = 1, 2, \ldots, s$$

and

$$\text{(ii)} \quad \max_{j=1,2,\ldots,s} |\operatorname{Re}(\lambda_j)| \gg \min_{j=1,2,\ldots,s} |\operatorname{Re}(\lambda_j)|,$$

where $\lambda_j, j = 1, 2, \ldots, s$, are the eigenvalues of $\mathbf{A}$. These are assumed to be distinct. The non-linear system (1.5) with a given initial condition $\mathbf{y}(x_0) = \mathbf{y}_0$ may be regarded as being stiff in an interval I of x if the eigenvalues of the Jacobian matrix $\dfrac{\partial \mathbf{f}(x, \mathbf{y}(x))}{\partial \mathbf{y}}$, where $\mathbf{y}(x)$ is the solution of (1.5) and

$$\left[\frac{\partial \mathbf{f}(x, \mathbf{y}(x))}{\partial \mathbf{y}}\right]_{i,j} \equiv \frac{\partial f_i(x, \mathbf{y}(x))}{\partial y_j}$$

behave in a similar fashion in I. If we wish to integrate a stiff system of equations, it is necessary to ensure that our integration procedure possesses some adequate region of absolute stability. Probably the most famous stability requirement in this context is that of *A-stability* proposed by Dahlquist (1963).

Definition 1.5

A numerical method is said to be A-stable if its region of absolute stability contains the whole of the left hand half plane $\text{Re}(q) < 0$.

The requirement that an integration procedure should be A-stable imposes a severe restriction on the order of acceptable linear multistep methods, as the following theorem due to Dahlquist shows.

Theorem 1.2

(*i*) *An explicit linear multistep method cannot be A-stable.*

(*ii*) *The order of an A-stable implicit linear multistep method cannot exceed* 2.

Since in the case of linear multistep methods A-stability implies implicitness, we are faced at each step with the problem of having to solve a non-linear system of algebraic equations of the form

$$\mathbf{y}_{n+k} = h\beta_k \mathbf{f}_{n+k} + \mathbf{g}_n \tag{1.10}$$

for the required solution $\mathbf{y}_{n+k}$, where $\mathbf{g}_n$ is a known s-dimensional vector. Simple predictor–corrector iteration schemes of the form

$$\mathbf{y}_{n+k}^{(p+1)} = h\beta_k \mathbf{f}(x_{n+k}, \mathbf{y}_{n+k}^{(p)}) + \mathbf{g}_n, \tag{1.11}$$

where $\mathbf{y}_{n+k}^{(0)}$ is obtained using an explicit method, are not efficient for use with stiff systems of equations since the restriction that needs to be imposed on h in order to guarantee convergence is excessively severe. (In fact often as severe as the restriction imposed on h by stability considerations when an explicit method is used.) In view of this it is necessary to use a much more sophisticated iteration scheme than (1.11), such as some form of Newton iteration, in order that the required solution may be generated efficiently.

Severe as the restriction of A-stability is, in one sense it is not severe enough. If, for example, we apply the trapezoidal rule

$$y_{n+1} - y_n = (h/2)\{f_{n+1} + f_n\},$$

which is well known to be A-stable, to the scalar test equation (1.8), we obtain the relation

$$y_{n+1} = [(1 + q/2)/(1 - q/2)]y_n, \qquad q = h\lambda.$$

As $|q| \to \infty$ we have $y_{n+1}/y_n \to -1$ and so, instead of the solution being represented as a rapidly decreasing quantity, which is the true behaviour for very large $-q$, it is represented as a slowly decaying oscillatory solution. In order to overcome this problem it is necessary to impose an even more severe restriction on (1.6) than that of A-stability. One such restriction is that of *L-stability* which was originally proposed by Ehle (1969) and which may be defined in the following way.

Definition 1.6

A numerical method is said to be L-stable if it is A-stable and, in addition, when applied to the scalar test equation (1.8) all solutions tend to zero as $\text{Re}(q) \to -\infty$.

As mentioned previously the condition of A-stability serves to ensure that, when a numerical method is applied to (1.8), as n increases all spurious solutions introduced by errors committed at each step point decrease everywhere in the complex left hand half plane. It does not, however, ensure that the spurious solutions are decreasing faster than the required solution. We now define *dominant A-stability.* This is a stability restriction corresponding to the demand that our required solution is the dominant solution of the resulting recurrence relation and is a more severe restriction than that of A-stability.

Definition 1.7

The linear multistep method (1.6) is said to be dominantly A-stable if all the roots r_i of the stability polynomial satisfy $|r_i| < |r_1| < 1, i = 2, 3, \ldots, k$, for all $q = h\lambda$ lying in the complex left hand half plane.

It must again be emphasised that this definition only has meaning if we are able to define r_1 uniquely for all q. This is not always the case especially for large $|q|$. We now guarantee the correct asymptotic behaviour of our scheme for large values of $-q$ by also imposing the following restriction of *dominant L-stability.*

Definition 1.8

The linear multistep method (1.6) is said to be dominantly L-stable if it is dominantly A-stable and if, in addition, all roots of the stability polynomial satisfy

$$\lim_{\text{Re}(q) \to -\infty} r_i = 0 \quad \text{for} \quad i = 1, 2, \ldots, k.$$

It would seem from the stability point of view that any numerical method satisfying the condition of dominant L-stability would be ideally suited for the numerical solution of stiff systems of equations. However, it may well be that dominant L-stability is too severe a restriction and that ordinary L-stability is sufficient for practical purposes. The precise stability criterion required is as yet not definitely known and further experience should indicate precisely what sort of stability conditions are most suitable. Finally on this point we note that it is possible to define slightly less severe restrictions such as *dominant $A(\alpha)$-stability* and *dominant stiff stability*. These are derived from the concept of dominant A-stability in exactly the same way as $A(\alpha)$- and stiff stability are derived from A-stability.

Many attempts have been made to circumvent the severe restrictions imposed by Dahlquist's theorems. The approach here is either not to demand full A-stability but rather some lesser stability requirement (Gear, 1969; Widlund, 1967), or to consider a class of discretisation algorithms based on a non-polynomial interpolant (Lambert, 1974). Another possible approach which does not seem to have received much specific attention in the literature is that whereby the required solution of (1.6) is generated using a non-direct method. Dahlquist has proved his classic Theorem 1.2 only for the case where the required solution of the linear multistep method being used is found with a direct method. If this fundamental assumption is violated, the conclusions stated in Theorem 1.2 may well be invalidated. The first part of Dahlquist's theorem, which demands that the linear multistep method should be implicit, does not seem to be affected by this new approach. It will however be shown later that it is possible to develop A-stable implicit methods of order greater than two based on standard discretisation methods if non-direct methods of solution are used. It was mentioned in Section 1.2 that the required solution of our discretisation scheme can often be generated in a stable manner, even in the presence of unwanted increasing solutions, providing the original initial value problem can be replaced by an equivalent well-posed boundary value problem. By the word "equivalent" here we mean that the required solution of our initial value problem is also a solution of the re-posed boundary value problem over some specified interval. It is important to note that these two problems are not computationally equivalent. If this was so there would, of course, be no point in making the transformation. Schemes for solving stiff systems of equations are essentially iterative in general as a result of one of the conclusions of Dahlquist's theorems. Thus it is fairly natural to ask whether the stability properties of (1.6) could be improved if the required solution were to be generated from the outset using some kind of iterative method. One possible scheme for the solution of (1.6) could take the form

$$\alpha_m \mathbf{y}_{n+m}^{(t)} - h\beta_m \mathbf{f}(x_{n+m}, \mathbf{y}_{n+m}^{(t)}) = \sum_{j=0}^{m-1} (h\beta_j \mathbf{f}(x_{n+j}, \mathbf{y}_{n+j}^{(t)}) - \alpha_j \mathbf{y}_{n+j}^{(t)})$$

$$+ \sum_{j=m+1}^{k} (h\beta_j \mathbf{f}(x_{n+j}, \mathbf{y}_{n+j}^{(t-1)}) - \alpha_j \mathbf{y}_{n+j}^{(t-1)}), \qquad (1.12)$$

where $m \in (0, k)$. This iteration scheme requires only m initial conditions of the form $\mathbf{y}_i^{(t)} = \mathbf{y}_i, i = 0, 1, \ldots, m-1, t > 0$, and as will be seen later the other $k - m$ conditions necessary for the required solution to be determined uniquely will be specified outside the region of interest. If a sequence of approximations $\{\mathbf{y}_n^{(0)}\}$ is available for all $n > m$, the scheme (1.12) with respectively $t = 1, 2, \ldots$ may be used to obtain sequences $\{\mathbf{y}_n^{(t)}\}$ for each value of n in the range of interest. Under certain circumstances these sequences may converge to the required solution $\mathbf{y}_n$ as t increases, in the sense that

$$\lim_{t \to \infty} \mathbf{y}_n^{(t)} = \mathbf{y}_n \quad \text{for all } n = m, m+1, \ldots$$

It now seems clear that it is not necessary for the required solution of the linear multistep method to necessarily be the dominant solution and, in addition, that not all the solutions of (1.6) need to be decreasing. We shall demonstrate later that this is indeed the case and that it is possible to develop efficient procedures for the numerical solution of stiff systems of equations and these procedures will be based on discretisation algorithms which are not A-stable when solved directly but which exhibit A-stability when solved using a precisely defined iterative technique.

The use of iteration scheme (1.12) represents just one possible way in which the original initial value problem (1.6) can be transformed into a boundary value problem. In later chapters we shall analyse in detail this type of iteration scheme, and related schemes, in the hope of finding efficient numerical methods for the solution of initial value problems of the form (1.5). Our aim is to replace the original initial value problem by an equivalent boundary value problem. Thus it is to be expected that there are close connections between this approach and the approach currently in use for the generation of non-dominant solutions of linear recurrence relations outlined briefly in Section 1.2. As emphasised earlier we shall examine these relationships in some detail and exploit them as much as possible. In view of this we start off our main analysis in the next section with a survey of some of the more efficient procedures for generating the solutions of recurrence relations.

2

An Investigation of Algorithms for the Solution of Recurrence Relations

In this chapter we examine in detail some of the more competitive algorithms for the numerical solution of scalar linear recurrence relations with a view to extending them to both the non-linear and the vector cases. As will be seen most standard methods for the solution of linear problems cannot be extended directly to the non-linear case. It therefore seems necessary that some sort of iterative approach be adopted for the solution of non-linear problems. This motivates the development of a class of algorithms which, although generally not as efficient as existing ones for the solution of linear recurrence relations, may nevertheless be extended directly to the non-linear case. Due mainly to the existence of a very powerful algorithm proposed by Olver (1967), the situation regarding the numerical solution of second order linear recurrence relations is fairly satisfactory, as far as computational algorithms are concerned. However, the same cannot be said for higher order equations, and thus in this chapter the second order and higher order cases will be considered separately.

2.1 Second order linear recurrence relations

This section is concerned with the numerical solution of the second order, linear, inhomogeneous recurrence relation

$$a_2(n)y_{n+2} + a_1(n)y_{n+1} + a_0(n)y_n = g(n), \tag{2.1}$$

where $a_2(n)a_0(n) \neq 0$ for any n in the range of interest, together with the initial conditions $y_0 = k_0$, $y_1 = k_1$ which uniquely define the required solution of (2.1). For the time being it will be assumed that $y_n, g(n)$ and the $a_i(n)$ are sequences of scalars, either real or complex. However, later on some consideration will be given to the case where $g(n)$ and y_n are sequences of vectors with the $a_i(n)$ being sequences of matrices. The general solution of (2.1) may be written in the form

$$y_n = c_1 y_{1,n} + c_2 y_{2,n} + p_n, \tag{2.2}$$

where $y_{1,n}$ and $y_{2,n}$ are two linearly independent solutions of the homogeneous part of (2.1) and p_n is a particular solution. It will be assumed that $y_{2,n}$ may be chosen so that $y_{2,0} \neq 0$, that $y_{1,n}$ is a dominant solution in the sense of (1.3) and that the given initial conditions define a non-dominant solution of (2.1), i.e. of the form (2.2) with $c_1 = 0$. The *initial value problem* defined by (2.1) together with the two given initial conditions may also be re-formulated as a *boundary value problem* satisfying

$$y_0 = k_0, \qquad \lim_{n\to\infty} \frac{y_n}{y_{1,n}} = 0. \tag{2.3}$$

It may be verified immediately that the required solution of this re-posed problem is given by

$$y_n = \left[\frac{k_0 - p_0}{y_{2,0}}\right] y_{2,n} + p_n \tag{2.4}$$

and that this is also the required solution of the problem defined by (2.1) and the two original initial conditions, which were assumed to be such that $c_1 = 0$ in (2.2).

The first really effective algorithm for the solution of the initial value problem (2.1) in the homogeneous case ($g(n) \equiv 0$), the famous *backward recurrence algorithm*, was proposed by Miller (1952) in connection with the tabulation of modified Bessel functions. Basically the approach adopted by Miller was first of all to replace the initial condition $y_1 = k_1$ by the two boundary conditions $y_{N+1} = 0, y_N = 1$, for some suitably large integer N, in an attempt to find a solution of (2.1) which satisfies the second of the conditions (2.3). By recurring in the direction of decreasing n with initial conditions $y_N^{[N]} = 1, y_{N+1}^{[N]} = 0$, a sequence $y_n^{[N]}$ may be obtained from the relation

$$y_n^{[N]} = -\frac{1}{a_0(n)}\{a_1(n)y_{n+1}^{[N]} + a_2(n)y_{n+2}^{[N]}\}$$

for all $0 \leq n \leq N-1$. Once the quantity $y_0^{[N]}$ is obtained, the sequence $y_n^{[N]}$ may be normalised by multiplying through by the scaling factor $y_0/y_0^{[N]}$, to produce a normalised sequence $\hat{y}_n^{[N]}$ which satisfies the given initial condition $\hat{y}_0^{[N]} = k_0$.

A further sequence of approximation may be obtained by taking another starting point, say $N + 5$, and carrying out the same process. This procedure is continued until two successive sequences of normalised approximations differ by less than a prescribed amount in some appropriate sense.

An obvious drawback to this approach is the absence of an efficient procedure for determining a suitable value of N before any solution values have been computed. If the solution is required to some fixed precision ε, whether relative or absolute, over a range $0 \leq n \leq L$, there will usually be an optimal value of the integer N, $\hat{N}$ say, at which the boundary conditions may be specified so that the required solution is obtained to the prescribed degree of precision with a minimum amount of computational effort. If the value of N chosen is less than $\hat{N}$, which is of course unknown initially, insufficient precision will normally be obtained. If the chosen value of N is much greater than $\hat{N}$ an excessive amount of computation will be required. In view of this it is highly desirable that a reasonable estimate of $\hat{N}$ should be obtained before any solution values are calculated. Continuing this line of attack Gautschi (1967) pointed out that, if the value of y_N is known in advance for any large value of N (i.e. via an asymptotic expansion for example), the original problem could be re-formulated as the solution of the tri-diagonal system of linear algebraic equations

$$\left.\begin{aligned}
a_1(0)y_1 + a_2(0)y_2 \qquad\qquad\qquad\qquad &= g(0) - a_0(0)k_0 \\
a_0(1)y_1 + a_1(1)y_2 + a_2(1)y_3 \qquad\qquad &= g(1) \\
a_0(2)y_2 + a_1(2)y_3 + a_2(2)y_4 &= g(2) \\
\cdot \quad \cdot \quad \cdot \quad \cdot \quad \cdot \quad \cdot \quad \cdot \quad \cdot \quad \cdot \quad \cdot \quad \cdot \quad & \\
\cdot \quad \cdot \quad \cdot \quad \cdot \quad \cdot \quad \cdot \quad \cdot \quad \cdot \quad \cdot \quad \cdot \quad \cdot \quad & \\
a_0(N-3)y_{N-3} + a_1(N-3)y_{N-2} + a_2(N-3)y_{N-1} &= g(N-3) \\
a_0(N-2)y_{N-2} + a_1(N-2)y_{N-1} &= g(N-2) - a_2(N-2)y_N
\end{aligned}\right\} \quad (2.5)$$

Gautschi did not pursue this idea any further, however, because of the difficulty in general of obtaining a suitable value for y_N. This approach was, however, further extended by Olver (1967), who observed that Miller's backward recurrence algorithm may be regarded merely as a procedure for generating a solution of the tri-diagonal system (2.5). We shall assume that the solution of (2.1), uniquely specified by the given initial conditions, is required in the range $0 \leq n \leq L$. Olver's basic approach is to set $y_N = 0$ for some $N > L$

and then to solve (2.5) by simple elimination followed by back substitution. If the solution obtained from (2.5) with boundary conditions $y_0 = k_0, y_N = 0$ is denoted by $y_n^{[N]}$, it can be shown under fairly general conditions that

$$\lim_{N \to \infty} y_n^{[N]} = y_n \quad \text{for all} \quad n \in [1, L],$$

providing y_n is a non-dominant solution of (2.1). The first stage of Olver's algorithm is to reduce the system (2.5) to upper triangular form. The first equation of (2.5) may be rewritten in the form

$$-p_1 y_2^{[N]} + p_2 y_1^{[N]} = e_1,$$

where

$$p_1 = 1, \quad p_2 = \frac{-a_1(0)}{a_2(0)}, \qquad e_1 = \frac{a_0(0)\,k_0 - g(0)}{a_2(0)}. \tag{2.6}$$

Eliminating $y_1^{[N]}$ between (2.6) and the second of (2.5), we obtain

$$-p_2 y_3^{[N]} + p_3 y_2^{[N]} = e_2,$$

where

$$p_3 = \frac{-a_1(1)}{a_2(1)} p_2 - \frac{a_0(1)}{a_2(1)} p_1,$$

$$e_2 = \frac{a_0(1)}{a_2(1)} e_1 - \frac{p_2 g(1)}{a_2(1)}. \tag{2.7}$$

Continuing this elimination process we obtain for general n

$$-p_n y_{n+1}^{[N]} + p_{n+1} y_n^{[N]} = e_n,$$

where

$$p_{n+1} = \frac{-a_1(n-1)}{a_2(n-1)} p_n - \frac{a_0(n-1)}{a_2(n-1)} p_{n-1},$$

$$e_n = \frac{a_0(n-1)}{a_2(n-1)} e_{n-1} - \frac{g(n-1)}{a_2(n-1)} p_n. \tag{2.8}$$

Note that the sequence p_n satisfies the homogeneous form of the difference equation. Since $y_N^{[N]} = 0$, the first of equations (2.8) for $n = N - 1$ becomes

$$y_{N-1}^{[N]} = \frac{e_{N-1}}{p_N}.$$

By using the first of equations (2.8) with successively $n = N-2, N-3, \ldots$ the sequence $y_n^{[N]}$ may be computed in the direction of decreasing n for all

$n \in [1, N-1]$. An analysis of the truncation error of this procedure allows us to predict the effect of changing from N to $N+1$ before any back substitution has been performed. As a result, the optimal value of N (i.e. $\hat{N}$) which gives the required degree of precision over the complete range $[0, L]$ may be estimated automatically. It is this facility for estimating the optimal value of N in advance, more than anything else, which makes Olver's algorithm superior to other algorithms for generating non-dominant solutions of (2.1). If we replace N by $N+1$ in the first of equations (2.8) we obtain

$$-p_n y_{n+1}^{[N+1]} + p_{n+1} y_n^{[N+1]} = e_n.$$

Subtracting (2.8) from this expression we obtain

$$\begin{aligned} y_n^{[N+1]} - y_n^{[N]} &= \frac{p_n}{p_{n+1}} [y_{n+1}^{[N+1]} - y_{n+1}^{[N]}], \\ &= \frac{p_n}{p_{n+2}} [y_{n+2}^{[N+1]} - y_{n+2}^{[N]}], \\ &\quad \cdot \quad \cdot \quad \cdot \quad \cdot \quad \cdot \quad \cdot \\ &\quad \cdot \quad \cdot \quad \cdot \quad \cdot \quad \cdot \quad \cdot \\ &= \frac{p_n}{p_N} \frac{e_N}{p_{N+1}}. \end{aligned} \tag{2.9}$$

Suppose that we wish to compute y_L to D decimal places for given values of the integers L and D. The recurrence relations for p_n and e_n are first applied for $n = 2, 3, \ldots, L, L+1, \ldots$ until a value of n is reached for which

$$|p_L e_n/(p_n p_{n+1})| < (1/2) \times 10^{-D}, \tag{2.10}$$

and then N is taken as n and finally $y_N^{[N]}$ is set equal to 0. For several numerical examples illustrating this algorithm and for a discussion of various extensions to the procedure which has just been described, the reader is referred to Olver's original papers (Olver, 1967; Olver and Sookne, 1972).

This boundary value approach has led to the formulation of a very powerful algorithm, which is in general competitive with other existing methods, for the generation of non-dominant solutions of (2.1). One of the main practical applications of Olver's algorithm is in the tabulation of special functions, many of which satisfy second order linear recurrence relations. However, the algorithm is also capable of extension to the vector case, which has applications in the solution of systems of linear ordinary differential equations. Although these extensions are of a fairly trivial nature, they will be considered in some detail since we shall need to refer back to them at a later stage.

2.2 An extension to the vector case

We now consider the vector linear recurrence relation

$$[\mathbf{a}_2(n)]\mathbf{y}_{n+2} + [\mathbf{a}_1(n)]\mathbf{y}_{n+1} + [\mathbf{a}_0(n)]\mathbf{y}_n = \mathbf{g}(n), \tag{2.11}$$

where $\mathbf{y}_n$ and $\mathbf{g}(n)$ are s-dimensional vectors and the $\mathbf{a}_i(n)$ are sequences of square matrices of order $s \times s$. For the purposes of the analysis it will be assumed that both $\mathbf{a}_0(n)$ and $\mathbf{a}_2(n)$ are non-singular for all n. A general theory for vector recurrence relations may be found in Miller (1968). In particular Miller shows that (2.11) has a fundamental set of matrix solutions $\mathbf{F}(n)$, $\mathbf{G}(n)$ which each satisfy the homogeneous form of (2.11) and which are such that

$$\det\begin{bmatrix} \mathbf{F}(n-1) & \mathbf{G}(n-1) \\ \mathbf{F}(n) & \mathbf{G}(n) \end{bmatrix} \neq 0 \quad \text{for all} \quad n.$$

If now $\mathbf{p}_n$ is a particular solution of (2.11), any solution of (2.11) may be written in the form

$$\mathbf{y}_n = \mathbf{F}(n)\mathbf{c}_1 + \mathbf{G}(n)\mathbf{c}_2 + \mathbf{p}_n \quad \text{for all } n, \tag{2.12}$$

where $\mathbf{c}_1$ and $\mathbf{c}_2$ are appropriate s-dimensional constant vectors. It will be assumed that it is possible to choose the fundamental set of matrix solutions such that $[\mathbf{F}(n)]^{-1}$ exists for all n and

$$\lim_{n\to\infty} \|[\mathbf{F}(n)]^{-1}\| \, \|\mathbf{G}(n)\| = 0. \tag{2.13a}$$

It will also be assumed that the required solution $\mathbf{y}_n$ of (2.11) is such that

$$\lim_{n\to\infty} \|([\mathbf{F}(n)]^{-1})\mathbf{y}_n\| = 0 \tag{2.13b}$$

and that in (2.12) the vector $\mathbf{c}_1 = 0$, the null s-dimensional vector. It is easy to show (R. G. Keys and A. C. Reynolds, Theory of vector difference equations and an application to parabolic problems, unpublished manuscript) that under these conditions the computation of the solution $\mathbf{y}_n$ using direct forward recurrence is an unstable process. In view of this it would be advantageous if Olver's algorithm could be extended to enable us to generate $\mathbf{y}_n$ in a stable manner in the forward direction. The following theorem regarding the convergence of a natural extension of Olver's algorithm to the vector case has been proved by Keys and Reynolds.

Theorem 2.1

Suppose that $\mathbf{F}(n)$ *and* $\mathbf{G}(n)$ *form a fundamental set of matrix solutions of the vector equation* (2.11). *Assume that* $\mathbf{G}(0)$ *is non-singular and that there exists an integer* J *such that* $\mathbf{F}(n)$ *is non-singular for all* $n \geq J$. *Suppose further that* (2.13a) *holds and* $\mathbf{y}_n$ *is a solution of* (2.11) *such that* (2.13b) *also holds. Let* N *be any integer greater than* J *and let*

$$\mathbf{y}_0^{[N]} = \mathbf{y}_0, \qquad \mathbf{y}_N^{[N]} = 0. \tag{2.13c}$$

Under these conditions if, for all integers $N \geq J$, *a unique solution* $\mathbf{y}_n^{[N]}$ *of* (2.11) *exists and satisfies the boundary conditions* (2.13c) *then*

$$\lim_{N\to\infty} \mathbf{y}_n^{[N]} = \mathbf{y}_n.$$

As a consequence of Theorem 2.1 it follows that the required solution of the original problem (2.11) may be approximated arbitrarily closely by the solution of a boundary value problem of the form (2.5) with appropriate vector-valued functions replacing the scalars. The procedure for computing the solution $\mathbf{y}_n$ is as follows:

Define the block tri-diagonal matrix $\mathbf{A}^{[N]}$ *as*

$$\mathbf{A}^{[N]} = \begin{bmatrix} \mathbf{a}_1(0) & \mathbf{a}_2(0) & & & & \\ \mathbf{a}_0(1) & \mathbf{a}_1(1) & \mathbf{a}_2(1) & & & \\ & \cdot & \cdot & \cdot & & \\ & & \cdot & \cdot & \cdot & \\ & & & \mathbf{a}_0(N-3) & \mathbf{a}_1(N-3) & \mathbf{a}_2(N-3) \\ & & & & \mathbf{a}_0(N-2) & \mathbf{a}_1(N-2) \end{bmatrix}$$

and the vector $\mathbf{b}^{[N]}$ *as*

$$\mathbf{b}^{[N]} = \begin{bmatrix} \mathbf{g}(0) - \mathbf{a}_0(0)\mathbf{y}_0 \\ \mathbf{g}(1) \\ \cdot \\ \cdot \\ \cdot \\ \mathbf{g}(N-2) - \mathbf{a}_2(N-2)\mathbf{y}_N \end{bmatrix}.$$

Then providing that $\mathbf{A}^{[N]}$ *is non-singular and N is sufficiently large, the solution* $\mathbf{y}^{[N]}$ *of*

$$\mathbf{A}^{[N]}\mathbf{y}^{[N]} = \mathbf{b}^{[N]}$$

gives the required approximation to $\mathbf{y}_n$ *over the given range.*

A procedure now needs to be derived for estimating in advance an optimal value of N. In order to simplify the notation it is assumed from now on that $\mathbf{a}_2(n) \equiv \mathbf{I}_s$, the $s \times s$ unit matrix. Systems for which $\mathbf{a}_2(n)$ is singular for some n, so that we are not able to multiply our equations through by $[\mathbf{a}_2(n)]^{-1}$, do not arise in the practical applications which we have in mind and so will not concern us any further although this problem could, no doubt, be overcome if it were to arise in a more general context. Following the approach adopted in the scalar case, the tri-diagonal system of equations corresponding to (2.5) may be reduced to the "upper triangular" form

$$\mathbf{p}_n\mathbf{y}_{n+1}^{[N]} - \mathbf{y}_n^{[N]} = \mathbf{e}_n, \qquad n \geq 1 \tag{2.14}$$

using simple forward elimination where

$$\begin{aligned} \mathbf{p}_1 &= -[\mathbf{a}_1(0)]^{-1}, \mathbf{e}_1 = -\mathbf{p}_1\{\mathbf{a}_0(0)\mathbf{k}_0 - \mathbf{g}(0)\} \\ \mathbf{p}_n &= -[\mathbf{a}_1(n-1) + \mathbf{a}_0(n-1)\mathbf{p}_{n-1}]^{-1} \\ \mathbf{e}_n &= \mathbf{p}_n\{\mathbf{a}_0(n-1)\mathbf{e}_{n-1} + \mathbf{g}(n-1)\} \end{aligned} \Bigg\} \quad n = 2, 3, \ldots \tag{2.15}$$

and where it is assumed that all the required inverses exist.

We emphasise here that the $\mathbf{p}_n$ are matrices and the $\mathbf{e}_n$ are vectors. Since we have set $\mathbf{y}_N^{[N]} = 0$, the final equation of (2.14) reduces to

$$\mathbf{y}_{N-1}^{[N]} = -\mathbf{e}_{N-1}$$

and the remaining values of $\mathbf{y}_n^{[N]}$ in the range $[1, N-2]$ may be found using simple back substitution in the usual way. Before any solution values have been obtained an optimal value of N may be estimated using the following extension of Olver's technique. From (2.14) with N replaced by $N+1$ we have

$$\mathbf{p}_n\mathbf{y}_{n+1}^{[N+1]} - \mathbf{y}_n^{[N+1]} = \mathbf{e}_n$$

and the subtraction of (2.14) from this relation produces

$$\begin{aligned} \mathbf{y}_n^{[N+1]} - \mathbf{y}_n^{[N]} &= \mathbf{p}_n[\mathbf{y}_{n+1}^{[N+1]} - \mathbf{y}_{n+1}^{[N]}], \\ &= \mathbf{p}_n\mathbf{p}_{n+1}[\mathbf{y}_{n+2}^{[N+1]} - \mathbf{y}_{n+2}^{[N]}], \\ &\cdots \\ &= \mathbf{p}_n\mathbf{p}_{n+1}\cdots\mathbf{p}_{N-1}\mathbf{y}_N^{[N+1]}, \\ &= -\mathbf{p}_n\mathbf{p}_{n+1}\cdots\mathbf{p}_{N-1}\mathbf{e}_N. \end{aligned}$$

This expression for the truncation error which, as is to be expected, is much more complicated than in the scalar case can now be used to predict the effect of changing from N to $N + 1$ before any back substitution has been performed and this can be used to estimate the optimal value of N. If, for example, we wish to calculate $\mathbf{y}_L$ to D decimal places for given values of the integers L and D, we continue to perform the forward elimination for values of n past $n = L$ until a value of n is reached for which

$$\|\mathbf{p}_L\mathbf{p}_{L+1}\cdots\mathbf{p}_{n-1}\mathbf{e}_n\| < \tfrac{1}{2} \times 10^{-D}$$

and then this value of n is taken as N.

Example 2.1

We illustrate our algorithm by applying it to the solution of the system

$$\mathbf{y}_{n+2} + \begin{bmatrix} -12\tfrac{1}{2} & -1\tfrac{1}{2} \\ 3 & -8 \end{bmatrix}\mathbf{y}_{n+1} + \begin{bmatrix} 15\tfrac{1}{2} & 5\tfrac{1}{2} \\ -11 & -1 \end{bmatrix}\mathbf{y}_n = 0$$

with the initial condition $\mathbf{y}_0 = \begin{bmatrix} 0 \\ 1 \end{bmatrix}$ and with the additional information that a non-dominant solution of this system is required.

The general solution of this system may be written in the form

$$\mathbf{y}_n = \begin{bmatrix} 10^n & -9^n \\ -10^n & 2\times 9^n \end{bmatrix}\begin{bmatrix} c_{11} \\ c_{12} \end{bmatrix} + \begin{bmatrix} 1 & -(\tfrac{1}{2})^n \\ -1 & 2\times(\tfrac{1}{2})^n \end{bmatrix}\begin{bmatrix} c_{21} \\ c_{22} \end{bmatrix}.$$

It is easy to verify that the boundary conditions imposed on the solution are such that $c_{11} = c_{12} = 0$, $c_{21} = c_{22} = 1$.

The matrices $\mathbf{F}(n)$ and $\mathbf{G}(n)$ and the required solution $\mathbf{y}_n$ all of which appeared earlier in (2.12) may be identified as

$$\mathbf{F}(n) = \begin{bmatrix} 10^n & -9^n \\ -10^n & 2\times 9^n \end{bmatrix}, \mathbf{G}(n) = \begin{bmatrix} 1 & -(\tfrac{1}{2})^n \\ -1 & 2\times(\tfrac{1}{2})^n \end{bmatrix},$$

$$\mathbf{y}_n = \begin{bmatrix} 1 - (\tfrac{1}{2})^n \\ -1 + 2 \times (\tfrac{1}{2})^n \end{bmatrix},$$

from which it can be seen that $\mathbf{F}(n)$ and $\mathbf{G}(n)$ satisfy (2.13a) whilst $\mathbf{y}_n$ satisfies (2.13b). The particular problem which is considered is that of computing the required solution $\mathbf{y}_n$ correct to six decimal places at $n = 11$ and the results obtained are given in Table 2.1.

TABLE 2.1

n	Estimated truncation error in $y_{11}^{[n]}$		$y_{n-1}^{[18]}$		Error in $y_{n-1}^{[18]}$	
12	$-0{\cdot}899744 \times 10^{-1}$	$0{\cdot}899488 \times 10^{-1}$	0·9995116	−0·9990233	$0{\cdot}9999922 \times 10^{-7}$	$0{\cdot}9999844 \times 10^{-7}$
13	$-0{\cdot}899858 \times 10^{-2}$	$0{\cdot}899715 \times 10^{-2}$	0·9997548	−0·9995107	$0{\cdot}9999928 \times 10^{-6}$	$0{\cdot}9999857 \times 10^{-6}$
14	$-0{\cdot}899921 \times 10^{-3}$	$0{\cdot}899842 \times 10^{-3}$	0·9998679	−0·9997458	$0{\cdot}9999935 \times 10^{-5}$	$0{\cdot}9999871 \times 10^{-5}$
15	$-0{\cdot}899956 \times 10^{-4}$	$0{\cdot}899912 \times 10^{-4}$	0·9998389	−0·9997779	$0{\cdot}9999942 \times 10^{-4}$	$0{\cdot}9999884 \times 10^{-4}$
16	$-0{\cdot}899976 \times 10^{-5}$	$0{\cdot}899951 \times 10^{-5}$	0·9989695	−0·9989390	$0{\cdot}9999948 \times 10^{-3}$	$0{\cdot}9999895 \times 10^{-3}$
17	$-0{\cdot}899986 \times 10^{-6}$	$0{\cdot}899973 \times 10^{-6}$	0·9899847	−0·9899696	$0{\cdot}9999953 \times 10^{-2}$	$0{\cdot}9999906 \times 10^{-2}$
18	$-0{\cdot}899992 \times 10^{-7}$	$0{\cdot}899985 \times 10^{-7}$	0·8999927	−0·8999856	$0{\cdot}9999958 \times 10^{-1}$	$0{\cdot}9999915 \times 10^{-1}$

From Table 2.1 it can be seen immediately that the estimated truncation errors suggest that a value $N = 18$ should be taken. The values of $\mathbf{y}_n^{[18]}$ obtained for the range [11, 17] are given in the table, from which it can be seen that the value $\mathbf{y}_{11}^{[18]}$ agrees with the true solution to the required six decimal places.

Olver's basic approach, although very effective for the numerical solution of second order linear recurrence relations, may not be extended immediately to non-linear and higher order linear recurrence relations mainly on account of the following two difficulties:

(1) Olver's algorithm for the determination of the optimal value of N has not been generally extended to linear equations of order greater than two.

(2) By virtue of the fact that Olver's algorithm is inherently based on the solution of a system of linear algebraic equations using Gaussian elimination, it would seem to be difficult to extend it directly to deal with classes of non-linear recurrence relations.

Since we shall be mainly interested in later chapters in the numerical solution of non-linear recurrence relations arising from the solution of ordinary differential equations using finite difference techniques, our main concern now is in overcoming problem (2). If a successful attempt is to be made to extend our boundary value approach to non-linear recurrence relations, it would seem that some sort of iterative scheme will be needed for the solution of the non-linear algebraic equations. With this problem in mind we shall first of all derive an iterative algorithm for the solution of the linear system (2.1). Although not in general as efficient as Olver's algorithm for the solution of linear problems, this iterative algorithm has the important property that it can be extended directly to a large class of non-linear problems. An interesting link between this iterative scheme and the Olver algorithm for the solution of second order linear recurrence relations will be derived. At present we shall not be interested in the derivation of the iteration scheme and this will be investigated more fully in Section 2.5. Instead the scheme will be produced "out of the blue" and an analysis of some of its basic properties, particularly its relation to Olver's algorithm, will be given.

2.3 Iterative algorithms for the solution of linear recurrence relations

For the numerical solution of (2.1) we now consider the particular class of iteration schemes defined by

$$\begin{aligned} a_2(n)y_{n+2}^{(t-1)} + b(n)y_{n+1}^{(t)} + a_0(n)y_n^{(t)} \\ = g(n) + \{b(n) - a_1(n)\}y_{n+1}^{(t-1)}, \end{aligned} \tag{2.16}$$

where at this stage $b(n)$ is an arbitrary function of n. It is clear that if a sequence of scalars $b(n)$ can be found which causes the scheme (2.16) to converge, then the convergence is to a solution of (2.1) since (2.16) reduces identically to (2.1) on convergence. We first consider the particular scheme in which the sequence $b(n)$ is defined by the relations

$$\left.\begin{aligned} b(n) &= a_1(n) - \frac{a_0(n)a_2(n-1)}{b(n-1)}, \qquad n = 0, 1, 2, \ldots, \\ b(-1) &= \infty. \end{aligned}\right\} \tag{2.17}$$

We will show how this scheme may be derived in Section 2.5 but for the time being it will be produced without further comment. The only initial condition necessary for this scheme is that $y_0^{(t)} = k_0$ for all integers $t \geqslant 0$. The sequence of initial approximations $\{y_n^{(0)}\}$ to $\{y_n\}$ for all $n = 1, 2, \ldots$ allows (2.16), with increasing integer values of t, to be used to generate sequences $\{y_n^{(t)}\}$, which in certain circumstances will converge to the required solution $\{y_n\}$ as t increases.

We now consider (2.1) as having arisen from the numerical solution of a linear first order ordinary differential equation using a linear two-step method. In this case the desired procedure, especially when integrating stiff systems of equations, will in general be to advance one integration step at a time. On the basis of an estimate obtained for the local truncation error of the particular scheme being used, a decision relating to the validity of the current steplength of integration can be made. By this we mean that at each step a decision needs to be made relating to whether the current integration step is: too small and so may be increased while still giving the required degree of precision; about right so the solution may be accepted and the step may remain unchanged; or is too large and so the solution is unacceptable. In view of this, when integrating from the ith step point, x_i say, we shall wish to compute our final iterate at the next step point x_{i+1} as quickly as possible; i.e. before too many iterates at steppoints x_j, $j > i + 1$, have been computed. Another important practical consideration which must be borne in mind when using iterative schemes of this type is the need to keep the amount of storage space required to a minimum, especially if we wish to integrate large systems of equations over long intervals of x. In view of this any integration procedure which is such that $y_n^{(1)}$ is calculated first of all for all n in the range of interest, then $y_n^{(2)}$ is calculated for all n, and so on until convergence will not normally be a practical proposition. As a result of the stepsize and storage considerations the sequence of iterates will in fact be computed in the order shown in Fig. 2.1.

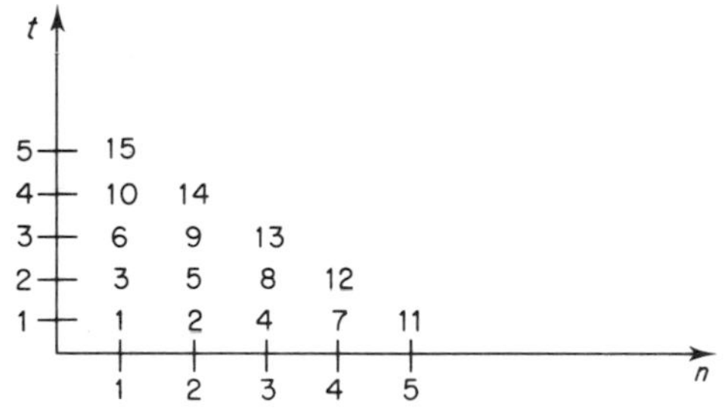

FIG. 2.1 Order of computation of iterates for (2.16)

Thus, for example, $y_3^{(1)}$ is the fourth iterate to be calculated and $y_2^{(3)}$ is the ninth. From Fig. 2.1 it can be seen that our iteration scheme is carried out along diagonals $n + t = k$ for $k = 2, 3, \ldots$. In order to derive a formal relationship between this iteration scheme and Olver's algorithm we first need to prove the following lemma.

Lemma 2.1

Let the sequence $b(n)$ be defined as in (2.17) and let the sequences p_n and e_n be defined as in (2.8). Then

$$\text{(a)} \quad b(n) = \frac{-p_{n+2}}{p_{n+1}} a_2(n) \quad \text{for all } n \geq 0,$$

$$\text{(b)} \quad \frac{e_{N-j-1}}{p_{N-j}} = \frac{1}{b(N-j-2)} \{-a_0(N-j-2)y_{N-j-2}^{(\alpha+1)} +$$

$$[b(N-j-2) - a_1(N-j-2)]y_{N-j-1}^{(\alpha)} + g(N-j-2)\}$$

for all $N > 0$, $j < N - 1$ and for all integers $\alpha \geq 0$.

Proof
The proofs of both parts of this lemma are by induction. We consider first of all

Part (a)
For $n = 0$ we have

$$\frac{-p_2}{p_1} a_2(0) = \frac{a_1(0)}{a_2(0)} a_2(0) = a_1(0) = b(0).$$

Assuming that the result is valid for $n = k - 1$ we have

$$\frac{-p_{k+2}}{p_{k+1}} a_2(k) = \frac{p_{k+1}a_1(k) + p_k a_0(k)}{p_{k+1}} = a_1(k) + \frac{p_k}{p_{k+1}} a_0(k),$$

$$= a_1(k) - \frac{a_0(k)a_2(k-1)}{b(k-1)},$$

$$= b(k).$$

By induction the result (a) follows immediately.

Part (b)
It is assumed that result (b) holds for $j = N - 2$ and on the assumption that it holds for $j = i + 1$ it will be proved to be valid for $j = i$. An inductive argument will then show that the result is valid for all $j < N - 1$. On setting $j = N - 2$ in

(b) we have

$$\text{L.H.S.} = \frac{e_1}{p_2} = \frac{g(0) - a_0(0)k_0}{a_1(0)} = \text{R.H.S.}$$

For $j = i$ and using (2.16), (2.17) and (2.8) we have that

$$\text{R.H.S.} = \frac{1}{b(N-i-2)}\{-a_0(N-i-2)y_{N-i-2}^{(\alpha+1)}$$

$$+ [b(N-i-2) - a_1(N-i-2)]y_{N-i-1}^{(\alpha)} + g(N-i-2)\},$$

$$= \frac{-a_0(N-i-2)}{b(N-i-2)b(N-i-3)}\{[b(N-i-3)$$

$$- a_1(N-i-3)]y_{N-i-2}^{(\alpha)} - a_0(N-i-3)y_{N-i-3}^{(\alpha+1)}$$

$$- a_2(N-i-3)y_{N-i-1}^{(\alpha)} + g(N-i-3)\}$$

$$+ \frac{1}{b(N-i-2)}\{[b(N-i-2) - a_1(N-i-2)]y_{N-i-1}^{(\alpha)}$$

$$+ g(N-i-2)\},$$

$$= \frac{-a_0(N-i-2)}{b(N-i-2)b(N-i-3)}\{[b(N-i-3)$$

$$- a_1(N-i-3)]y_{N-i-2}^{(\alpha)}$$

$$-a_0(N-i-3)y_{N-i-3}^{(\alpha+1)} + g(N-i-3)\} + \frac{g(N-i-2)}{b(N-i-2)},$$

$$= \frac{-a_0(N-i-2)}{b(N-i-2)}\frac{e_{N-i-2}}{p_{N-i-1}} + \frac{g(N-i-2)}{b(N-i-2)}$$

by induction hypothesis,

$$= -\left\{g(N-i-2) - \frac{a_0(N-i-2)e_{N-i-2}}{p_{N-i-1}}\right\}\frac{p_{N-i-1}}{a_2(N-i-2)p_{N-i}}$$

from part (a),

$$= \frac{1}{p_{N-i}}\left\{\frac{a_0(N-i-2)e_{N-i-2}}{a_2(N-i-2)} - p_{N-i-1}\frac{g(N-i-2)}{a_2(N-i-2)}\right\},$$

$$= \frac{e_{N-i-1}}{p_{N-i}}.$$

This completes the proof of the lemma.

The following theorem can now be proved.

Theorem 2.2

Denote the solution of the linear system (2.5) *by* $y_n^{[N]}$ *and the solution of the iteration scheme* (2.16), (2.17) *with initial approximations* $y_n^{(0)} = 0$ *for all* $n > 0$, $y_0^{(t)} = y_0$ *for all* $t \geq 0$ *by* $y_n^{(t)}$ *then*

$$y_n^{[N]} = y_n^{(N-n)}. \tag{2.18}$$

Proof

The proof of this theorem will be by induction.

Step 1

We first of all prove that

$$y_{N-1}^{(1)} = y_{N-1}^{[N]} \tag{2.18a}$$

for all $N \geq 2$ by induction on N. For $N = 2$ we have

$$y_1^{(1)} = \frac{g(0) - a_0(0)y_0}{a_1(0)} = y_1^{[2]}.$$

Suppose now that we assume the result (2.18a) to be valid for $N = k$, i.e. $y_{k-1}^{(1)} = y_{k-1}^{[k]}$. We now wish to establish the result for $N = k + 1$. From (2.16) with $n = k - 1$ and $t = 1$ we have

$$y_k^{(1)} = \frac{g(k-1) - a_0(k-1)y_{k-1}^{(1)}}{b(k-1)},$$

$$= \frac{g(k-1) - a_0(k-1)y_{k-1}^{[k]}}{b(k-1)} \quad \text{by induction hypothesis,}$$

$$= \frac{g(k-1) - a_0(k-1)y_{k-1}^{[k]}}{-\left(\dfrac{p_{k+1}}{p_k}\right)a_2(k-1)} \quad \text{from lemma (2.1)(a),}$$

$$= \frac{a_0(k-1)y_{k-1}^{[k]}p_k - g(k-1)p_k}{p_{k+1}a_2(k-1)},$$

$$= \frac{a_0(k-1)e_{k-1} - p_k g(k-1)}{p_{k+1}a_2(k-1)} \quad \text{since } y_{k-1}^{[k]} = \frac{e_{k-1}}{p_k},$$

$$= \frac{e_k}{p_{k+1}} \quad \text{from (2.8),}$$

$$= y_k^{[k+1]} \quad \text{from the first of (2.8) with } N = k+1, n = k.$$

This establishes the proof of the theorem for $n = N - 1$.

Step 2

We now assume that the theorem holds for $n = N - i$ and prove that it is valid for $n = N - i - 1$, where $i < N - 1$. By our induction hypothesis it follows that $y_{N-i}^{[N]} = y_{N-i}^{(i)}$ and we wish to prove that $y_{N-i-1}^{[N]} = y_{N-i-1}^{(i+1)}$. From (2.8) we have that

$$-p_{N-i-1}y_{N-i}^{[N]} + p_{N-i}y_{N-i-1}^{[N]} = e_{N-i-1} \qquad \text{and so}$$

$$y_{N-i-1}^{[N]} = \frac{e_{N-i-1} + p_{N-i-1}y_{N-i}^{[N]}}{p_{N-i}},$$

$$= \frac{1}{b(N-i-2)}\{-a_0(N-i-2)y_{N-i-2}^{(i+1)}$$

$$+ [b(N-i-2) - a_1(N-i-2)]y_{N-i-1}^{(i)} + g(N-i-2)$$

$$- a_2(N-i-2)y_{N-i}^{(i)}\} \qquad \text{from Lemma 2.1 with } \alpha = i,$$

$$= y_{N-i-1}^{(i+1)} \qquad \text{from (2.16).}$$

This completes the proof of the theorem.

Theorem 2.2 serves to establish an interesting link between two seemingly different approaches to the numerical solution of (2.1). It is to be expected that for the linear case the iteration scheme defined by (2.16), (2.17) will not be as efficient as Olver's algorithm, or similar algorithms for the solution of higher order equations such as the one developed by Oliver (1968b), since it does not attempt to estimate in advance the optimal value of N but instead calculates a sequence of solutions with $N = 2, 3, \ldots$ in succession. However, although not as efficient in the linear case this class of iteration schemes has the advantage that it may be extended in a straightforward manner to deal with certain non-linear equations.

It will be recalled that the particular sequence $b(n)$ defined in (2.17) was produced without justification. This choice for $b(n)$ is clearly only one of many and we thus consider others. An obvious choice is to set $b(n) \equiv a_1(n)$ to produce the much simplified iteration scheme for the case $g(n) \equiv 0$.

$$a_2(n)y_{n+2}^{(t-1)} + a_1(n)y_{n+1}^{(t)} + a_0(n)y_n^{(t)} = 0. \qquad (2.19a)$$

If we define two triangular matrices $\mathbf{L}_N$, $\mathbf{U}_N$ and a vector $\mathbf{b}_N$ as

$$-\mathbf{L}_N = \begin{bmatrix} 0 & & & & \\ a_0(1)/a_1(1) & 0 & & & \\ & a_0(2)/a_1(2) & 0 & & \\ & & \ddots & \ddots & \\ & & & -a_0(N-2)/a_1(N-2) & 0 \end{bmatrix} \quad -\mathbf{b}_N = \begin{bmatrix} a_0(0)y_0/a_1(0) \\ 0 \\ \vdots \\ 0 \\ a_2(N-2)y_N/a_1(N-2) \end{bmatrix}$$

$$-\mathbf{U}_N = \begin{bmatrix} 0 & a_2(0)/a_1(1) & & & \\ & 0 & a_2(1)/a_1(1) & & \\ & & \ddots & \ddots & \\ & & & 0 & -a_2(N-3)/a_1(N-3) \\ & & & & 0 \end{bmatrix}$$

we may write iteration scheme (2.19a) in the form

$$\mathbf{y}^{(t)} = \mathbf{L}_N\mathbf{y}^{(t)} + \mathbf{U}_N\mathbf{y}^{(t-1)} + \mathbf{b}_N, \tag{2.19b}$$

where $\mathbf{y}^{(t)} = (y_1^{(t)}, y_2^{(t)}, \ldots y_{N-1}^{(t)})^T$, and this is precisely the Gauss–Seidel iteration scheme for the solution of system (2.5) with $g(n) \equiv 0$. Rearranging (2.19b) we obtain the expression

$$\mathbf{y}^{(t)} = (\mathbf{I} - \mathbf{L}_N)^{-1}\,[\mathbf{U}_N\mathbf{y}^{(t-1)} + \mathbf{b}_N]$$

and using knowledge of Gauss–Seidel iteration we know that this scheme converges for any given starting vector $\mathbf{y}^{(0)}$ if the spectral radius of the matrix $(\mathbf{I} - \mathbf{L}_N)^{-1}\,\mathbf{U}_N$ is less than unity. Since our iteration scheme involves using values $N = 3, 4, \ldots$ in succession it is necessary to ensure that

$$\rho[(\mathbf{I} - \mathbf{L}_N)^{-1}\,\mathbf{U}_N] < 1 \quad \text{for all} \quad N = 3, 4, \ldots$$

where $\rho(H)$ denotes the spectral radius of the matrix H. Since it may be difficult in practical applications to determine the spectral radii of this sequence of matrices we find it convenient to give an alternative criterion which provides a sufficient condition for this scheme to converge.

Lemma 2.2

Let

$$r_N = \max_{n \leqslant N}\left\{\left|\frac{a_0(n)}{a_1(n)}\right| + \left|\frac{a_2(n)}{a_1(n)}\right|\right\}.$$

Then if $r_N < 1$ for all N, scheme (2.19a) converges for any initial iterate with the ratio of the maximum norms of two successive error vectors being less than or equal to r_N.

The proof of a more general form of this lemma may be found in Isaacson and Keller (1966, p. 67). The lemma itself will be shown later to be of considerable importance in finding stability regions of linear multistep methods.

No further particular choices for the sequence $b(n)$ will be considered. Instead we now demonstrate our algorithm for a particular problem.

Example 2.2

Consider the solution of the second order, inhomogeneous recurrence relation

$$y_{n+1} - \frac{2n}{x}y_n + y_{n-1} = \frac{(\frac{1}{2}x)^n}{\Gamma(\frac{1}{2})\Gamma(n+\frac{3}{2})}$$

using the iterative scheme (2.16), (2.17). The particular solutions of this recurrence relation are the well-known Struve functions (Watson, 1952), which we denote by $\mathbf{H}_n(x)$. These are dominated by $\mathbf{Y}_n(x)$ in the direction of increasing n and by $\mathbf{J}_n(x)$ in the direction of decreasing n, where $\mathbf{Y}_n(x)$ and $\mathbf{J}_n(x)$ are the Bessel functions of the first kind and are two linearly independent solutions of the equation

$$y_{n+1} - \frac{2n}{x}y_n + y_{n-1} = 0.$$

The required solution is in effect the "middle" solution of our problem and so cannot be effectively generated either by recurring in the direction of increasing n or in the direction of decreasing n. In Table 2.2 we give the results obtained for the solution of our problem with $x = 0{\cdot}5$ and with initial conditions

$$y_0 = 0{\cdot}30955591, \qquad y_1 = 0{\cdot}05217365, \qquad y_n^{(0)} = 0 \qquad n \geq 2$$

in the range $2 \leq n \leq 10$, where the solution was required correct to eight significant figures. It was found that throughout the range four iterations produced the true solution to the required degree of precision.

TABLE 2.2

n	$y_n^{(1)}$	$y_n^{(2)}$	$y_n^{(3)}$	$y_n^{(4)}$
2	$0{\cdot}51954268 \times 10^{-2}$	$0{\cdot}52416513 \times 10^{-2}$	$0{\cdot}52423461 \times 10^{-2}$	$0{\cdot}52423576 \times 10^{-2}$
3	$0{\cdot}36979551 \times 10^{-3}$	$0{\cdot}37535427 \times 10^{-3}$	$0{\cdot}37544596 \times 10^{-3}$	$0{\cdot}37544751 \times 10^{-3}$
4	$0{\cdot}20480689 \times 10^{-4}$	$0{\cdot}20886133 \times 10^{-4}$	$0{\cdot}20893242 \times 10^{-4}$	$0{\cdot}20893364 \times 10^{-4}$
5	$0{\cdot}92834246 \times 10^{-6}$	$0{\cdot}95039536 \times 10^{-6}$	$0{\cdot}95079922 \times 10^{-6}$	$0{\cdot}95080623 \times 10^{-6}$
6	$0{\cdot}35613885 \times 10^{-7}$	$0{\cdot}36582100 \times 10^{-7}$	$0{\cdot}36600421 \times 10^{-7}$	$0{\cdot}36600742 \times 10^{-7}$
7	$0{\cdot}11842944 \times 10^{-8}$	$0{\cdot}12201148 \times 10^{-8}$	$0{\cdot}12208100 \times 10^{-8}$	$0{\cdot}12208223 \times 10^{-8}$
8	$0{\cdot}34754015 \times 10^{-10}$	$0{\cdot}35901921 \times 10^{-10}$	$0{\cdot}35924664 \times 10^{-10}$	$0{\cdot}35925069 \times 10^{-10}$
9	$0{\cdot}91263645 \times 10^{-12}$	$0{\cdot}94512511 \times 10^{-12}$	$0{\cdot}94577974 \times 10^{-12}$	$0{\cdot}94578144 \times 10^{-12}$
10	$0{\cdot}21685493 \times 10^{-13}$	$0{\cdot}22509472 \times 10^{-13}$	$0{\cdot}22526300 \times 10^{-13}$	$0{\cdot}22526619 \times 10^{-13}$

2.4 Non-linear recurrence relations

The main practical significance of the class of iterative algorithms developed in the previous section is the ease with which they may be extended to the solution of certain classes of non-linear recurrence relations. For the time being we shall confine our attention to the second order case and consider the recurrence relation

$$a_2(n)y_{n+2} + a_1(n)y_{n+1} + a_0(n)y_n = g(y_n, y_{n+1}, y_{n+2}) \tag{2.20}$$

which is of the type that can arise if an implicit linear two-step method is used to solve a non-linear ordinary differential equation. Following the approach adopted in the previous section for the linear case we could attempt to generate a solution of equation (2.20) using the iteration scheme

$$\begin{aligned} a_2(n)y_{n+2}^{(t-1)} &+ a_1(n)y_{n+1}^{(t)} + a_0(n)y_n^{(t)} \\ &= g(y_n^{(t)}, y_{n+1}^{(t-1)}, y_{n+2}^{(t-1)}). \end{aligned} \tag{2.21}$$

Since the required quantity $y_{n+1}^{(t)}$ is given explicitly in terms of known quantities if the sequence of iterates is calculated in the order defined by Fig. 2.1, this scheme does not present any computational difficulties which were not present in the linear case, providing of course that $a_1(n) \neq 0$ for any n. As we shall now show, scheme (2.21) may be regarded as a special case of the more general iteration scheme

$$a_2(n)y_{n+2}^{(t-1)} + a_1(n)y_{n+1}^{(t)} + a_0(n)y_n^{(t)}$$
$$= g(y_n^{(t)}, y_{n+1}^{(t)}, y_{n+2}^{(t-1)}), \tag{2.22}$$

which in turn may be regarded as a non-linear Gauss–Seidel iteration scheme. It follows immediately that if g is a non-linear function of $y_{n+1}^{(t)}$, then the required solution is defined implicitly by a set of non-linear algebraic equations. In order to solve these non-linear equations for $y_{n+1}^{(t)}$ it will be necessary for some standard iterative scheme to be used. This scheme will be referred to as the secondary iteration scheme whilst the non-linear Gauss–Seidel scheme will be called the primary iteration scheme. It will be shown that if one particular secondary iteration scheme is used then the complete algorithm reduces to (2.21), thus substantiating our claim that (2.22) is more general than (2.21). We now consider two particular secondary iteration schemes which are tailored to two specific practical applications. The first iteration scheme is that of direct iteration. If we rewrite scheme (2.22) in the form

$$a_1(n)y_{n+1}^{(t)} = g(y_n^{(t)}, y_{n+1}^{(t)}, y_{n+2}^{(t-1)}) + e(n), \tag{2.23}$$

where e is a known quantity, we may use the iteration scheme

$$a_1(n)y_{n+1,i+1}^{(t)}$$
$$= g(y_n^{(t)}, y_{n+1,i}^{(t)}, y_{n+2}^{(t-1)}) + e(n), \qquad \text{for } i = 0, 1, 2, \ldots, \tag{2.24}$$

where $y_{n+1,0}^{(t)} = y_{n+1}^{(t-1)}$. Putting

$$\phi(y_{n+1}^{(t)}) = \frac{1}{a_1(n)}\{g(y_n^{(t)}, y_{n+1}^{(t)}, y_{n+2}^{(t-1)}) + e(n)\}$$

we may rewrite iteration scheme (2.24) in the form

$$y_{n+1,i+1}^{(t)} = \phi(y_{n+1,i}^{(t)}). \tag{2.25}$$

Assuming that scheme (2.25) does converge, we may iterate with (2.25) for increasing i until convergence, in some appropriate sense, is obtained for $i = p$ say. At this point we then set $y_{n+1}^{(t)} = y_{n+1,\ p}^{(t)}$. If we use only $i = 0$, i.e. perform only one iteration cycle, we obtain precisely scheme (2.21). As a result the following theorem, the proof of which may be found in Henrici (1962), may be stated. This gives a sufficient condition for the convergence of (2.25) for iteration with increasing i.

Theorem 2.3

Let $\phi(y_n)$ satisfy the Lipschitz condition

$$|\phi(y_n) - \phi(y_n^*)| \leq M|y_n - y_n^*|$$

for all y_n, y_n^*, where the Lipschitz constant M satisfies $0 \leq M < 1$. Then there exists a unique solution $\alpha(n)$ of the equation $\phi(y_n^{(t)}) = y_n^{(t)}$ and furthermore

$$\lim_{i \to \infty} y_{n,i}^{(t)} = \alpha(n),$$

where $y_{n,i}^{(t)}$ is defined by (2.25) with $n + 1$ replaced by n.

This theorem may be extended directly to the vector case if, throughout, absolute values of scalars are replaced by norms of corresponding vectors. For the practical applications which we consider the function g in (2.24) is usually multiplied by a parameter h, which is a positive quantity denoting the steplength of integration. Thus when solving ordinary differential equations using linear multistep methods we can usually choose h to be sufficiently small to ensure convergence. In practice the direct iteration scheme is not usually iterated to convergence; in fact it is generally more efficient to perform just one direct iteration. Even so Theorem 2.3 provides a firm theoretical basis on which to found our scheme. Practical experience has shown that this procedure based on (2.24) is usually satisfactory for the solution of non-stiff systems of ordinary differential equations and we shall refer to it as the Gauss–Seidel – direct-iteration scheme. Since this procedure will mainly be of use in the numerical solution of systems of ordinary differential equations, we shall defer a more detailed discussion of its stability and convergence characteristics until the next chapter.

If we were now to think of scheme (2.20) as arising from the numerical solution of a *stiff* system of ordinary differential equations using an implicit linear two-step method, the Lipschitz constant M associated with ϕ will be very large unless h is excessively small. It follows that any scheme employing direct iteration will no longer converge to the required solution unless a very small value of h is used. It is this situation which we particularly wish to avoid when solving stiff systems of equations. This difficulty can be traced directly to the fact that, if direct iteration is used, our scheme is effectively explicit whereas it is known from Dahlquist's theorems that explicit linear multistep methods are of little use in the numerical solution of stiff systems. In view of this it is more appropriate, when dealing with stiff systems of equations, to replace the secondary direct iteration scheme by a more sophisticated scheme such as some form of Newton iteration. If we now rewrite the vector form of (2.22) as $\mathbf{f}(\mathbf{y}_{n+1}^{(t)}) = 0$, we could attempt to generate the required solution $\mathbf{y}_{n+1}^{(t)}$ using the straightforward Newton scheme

$$\mathbf{y}_{n+1,i}^{(t)} - \mathbf{y}_{n+1,i-1}^{(t)} = -\left[\frac{\partial \mathbf{f}}{\partial \mathbf{y}_{n+1}}(\mathbf{y}_{n+1,i-1}^{(t)})\right]^{-1} \mathbf{f}(y_{n+1,i-1}^{(t)}). \tag{2.26}$$

Given an initial approximation $\mathbf{y}_{n+1,0}^{(t)}$ this scheme may be iterated with increasing values of i until an end solution $\mathbf{y}_{n+1,m}^{(t)}$ say is obtained. (This solution will of course depend on the stopping criterion which is used.) Finally we set $\mathbf{y}_{n+1}^{(t)} = \mathbf{y}_{n+1,m}^{(t)}$. We shall refer to this scheme as a Gauss–Seidel–Newton scheme iterated to convergence. Although a procedure based on (2.26) is a perfectly feasible approach to the solution of the implicit equation (2.22) it will in general be rather an inefficient one due to the large number of iteration steps which will be required, especially if the primary iteration scheme converges slowly. When integrating stiff systems of equations we shall attempt to keep the total number of iterations which need to be performed to a minimum since each iteration step involves a function (and possibly a Jacobian) evaluation and these are normally computationally expensive. If we again consider scheme (2.22) we see immediately that in general there will be no point in iterating scheme (2.26) to convergence, because $\mathbf{y}_{n+1}^{(t)}$ will not in general be an accurate approximation to $\mathbf{y}_{n+1}$, especially for low values of t. This inaccuracy is due to errors in $\mathbf{y}_{n+2}^{(t-1)}$ and possibly $\mathbf{y}_n^{(t)}$ as approximations to $\mathbf{y}_{n+2}$ and $\mathbf{y}_n$ respectively. In this case practical experience has shown that it is more efficient in general if, when iterating for $\mathbf{y}_{n+1}^{(t)}$, just one (quasi-) Newton iteration is performed with $\mathbf{y}_{n+1,0}^{(t)} = \mathbf{y}_{n+1}^{(t-1)}$ and then we set $\mathbf{y}_{n+1}^{(t)} = \mathbf{y}_{n+1,1}^{(t)}$. Ortega and Rheinboldt (1970) refer to this scheme as the one-step Gauss–Seidel–Newton method and they conclude that it is one of the most efficient of all iteration schemes where Gauss–Seidel is the primary iteration; especially if the partial derivatives are not too difficult (computationally expensive) to evaluate. If we use scheme (2.26) to solve the non-linear system of equations (2.22) for the quantity $\mathbf{y}_{n+1}^{(t)}$ we obtain the composite iteration scheme

$$\left[\mathbf{a}_1(n) - \frac{\partial \mathbf{g}}{\partial \mathbf{y}_{n+1}}(\mathbf{y}_n^{(t)}, \mathbf{y}_{n+1}^{(t-1)}, \mathbf{y}_{n+2}^{(t-1)})\right]\left[\mathbf{y}_{n+1}^{(t)} - \mathbf{y}_{n+1}^{(t-1)}\right]$$
$$= -\{\mathbf{a}_2(n)\mathbf{y}_{n+2}^{(t-1)} + \mathbf{a}_1(n)\mathbf{y}_{n+1}^{(t-1)} + \mathbf{a}_0(n)\mathbf{y}_n^{(t)} - \mathbf{g}(\mathbf{y}_n^{(t)}, \mathbf{y}_{n+1}^{(t-1)}, \mathbf{y}_{n+2}^{(t-1)})\}. \tag{2.27}$$

It can now be seen that, although we are using a primary and a secondary iteration scheme which are both different, the composite iteration scheme (2.27) depends on the single iteration parameter t. The sequence of iterates can thus be calculated in the order defined by Fig. 2.1. Consideration of the stability and convergence properties of this scheme will again be deferred until Chapter 3, where the application of the scheme to the numerical solution of stiff systems of ordinary differential equations is considered. We may however state the following sufficient condition for the convergence of our iteration scheme (2.27) for the solution of the non-linear algebraic equation $\mathbf{Fy} = 0$. This condition may be regarded as a generalisation of that given in Lemma 2.2.

Theorem 2.4

Suppose that $\mathbf{F}:\mathbf{D} \subset \mathbb{R}^s \to \mathbb{R}^s$ *and that* $\mathbf{Fy}^* = 0$ *for some point* $\mathbf{y}^*$ *contained in the interior of* $\mathbf{D}$. *Suppose further that* $\mathbf{F}$ *is continuously differentiable in an open neighbourhood* $\mathbf{N}_0$ *of* $\mathbf{y}^*$ *where* $\mathbf{N}_0 \subset \mathbf{D}$. *Let* $\mathbf{D}(\mathbf{y})$, $-\mathbf{L}(\mathbf{y})$ *and* $-\mathbf{U}(\mathbf{y})$ *be the diagonal, strictly lower- and strictly upper-triangular parts of* $\mathbf{F}'(\mathbf{y})$ *and assume that* $\mathbf{D}(\mathbf{y}^*)$ *is non-singular. Define a matrix* $\mathbf{H}(\mathbf{y}^*)$ *as*

$$\mathbf{H}(\mathbf{y}^*) = [\mathbf{D}(\mathbf{y}^*) - \mathbf{L}(\mathbf{y}^*)]^{-1}[\mathbf{U}(\mathbf{y}^*)]$$

and assume that $\rho(\mathbf{H}(\mathbf{y}^*)) < 1$. *Then the Gauss–Seidel–Newton iteration scheme is well defined in an open ball of* $\mathbf{y}^*$ *with radius* $\mathbf{S}^*$ *and* $\mathbf{y}^*$ *is a* point of attraction *of the scheme. Here* $\mathbf{F}'$ *denotes the Frechet derivative of* $\mathbf{F}$.

The proof of this theorem may be found in Ortega and Rheinboldt (1970), p. 323.

Although the one-step Gauss–Seidel–Newton scheme (2.27) has proved to be efficient for practical applications it is by no means the only iteration scheme which is well suited to our problem. One obvious extension is to introduce a relaxation constant ω and then to use a successive overrelaxation scheme rather than the straightforward Gauss–Seidel scheme as the primary iteration. Hybrid schemes of this form were discussed in some detail by Ortega and Rheinboldt (1970), Section 7.4, but it is not clear whether they are in general more efficient than the straightforward Gauss–Seidel scheme $\omega = 1$ in this context. Ortega and Rheinboldt define a class of generalised linear methods of which the Gauss–Seidel scheme is one particular member and the distinguishing feature of the schemes in this class is that they consist of two distinct iterations—a primary and a secondary. Ortega and Rheinboldt reference many composite iteration schemes in which SOR is the primary iteration; examples are SOR–Newton, SOR–Steffenson. However, experience with these methods when used in conjunction with linear multistep methods for the solution of ordinary differential equations seems to be lacking at present.

Finally in this section we mention the possibility of solving non-linear algebraic equations of the form (2.20) in the much simpler case where g is a function only of y_n. Equations of this type may arise, for example, in the solution of non-stiff systems of ordinary differential equations using explicit linear multistep methods. In cases where we are not able to generate the required solution using a direct method, due to an unstable build-up of rounding errors, it may be possible to generate this solution using one of the iteration schemes mentioned previously. In the explicit case it is no longer necessary to use a secondary iteration scheme. Here Gauss–Seidel is used for the primary iteration and the iterates are computed in the order defined by Fig. 2.1. In order to illustrate our algorithm we consider a particular problem.

Example 2.3

Consider the solution of the non-linear recurrence relation

$$y_{n+1} - \left(\frac{n^2 + 5n + 10}{n + 1}\right) y_n + \left(\frac{2n^2 + 9n + 19}{4(n + 1)}\right) y_{n-1}$$
$$= \frac{(6n^2 + 25n + 55)}{36(n + 1)} 3^{n-1} y_{n-1}^2$$

with the boundary conditions $y_0 = 1$, $y_n \to 0$ as $n \to \infty$. We shall seek to generate the required solution of this equation using the Gauss–Seidel iteration scheme

$$y_{n+1}^{(t-1)} - \left(\frac{n^2 + 5n + 10}{n + 1}\right) y_n^{(t)} + \left(\frac{2n^2 + 9n + 19}{4(n + 1)}\right) y_{n-1}^{(t)}$$
$$= \left(\frac{6n^2 + 25n + 55}{36(n + 1)}\right) 3^{n-1} [y_{n-1}^{(t)}]^2.$$

It is difficult to determine *ab initio* whether or not this iteration scheme will converge with the initial approximations $y_n^{(0)} \equiv 0$, $n \geq 2$. Thus the approach adopted in practice is to "try it and see". This means that if the solution obtained satisfies the original recurrence relation, and has the correct asymptotic behaviour, it is accepted as the final solution. In Table 2.3a we give the solution obtained using a direct method starting from the analytic initial conditions $y_0 = 1, y_1 = 1/3$, rather than from the given boundary conditions. As can be seen there is a disastrous build-up of error rendering the direct method completely ineffective for the solution of our problem. In Table 2.3b we give the results obtained using the Gauss–Seidel iteration scheme together with the given boundary conditions. The required solution was sought correct to six significant figures and as can be seen this is achieved in five iterations.

TABLE 2.3a

n	y_n	True solution	n	y_n	True solution
2	0·11111111	0·11111111	11	$-0{\cdot}26428695 \times 10^{-4}$	$0{\cdot}56450292 \times 10^{-5}$
3	$0{\cdot}37037037 \times 10^{-1}$	$0{\cdot}37037037 \times 10^{-1}$	12	$-0{\cdot}48871001 \times 10^{-3}$	$0{\cdot}18816764 \times 10^{-5}$
4	$0{\cdot}12345679 \times 10^{-1}$	$0{\cdot}12345679 \times 10^{-1}$	13	$-0{\cdot}75117101 \times 10^{-2}$	$0{\cdot}62722547 \times 10^{-6}$
5	$0{\cdot}41152263 \times 10^{-2}$	$0{\cdot}41152263 \times 10^{-2}$	14	$-0{\cdot}22428501$	$0{\cdot}20907516 \times 10^{-6}$
6	$0{\cdot}13717420 \times 10^{-2}$	$0{\cdot}13717421 \times 10^{-2}$	15	$0{\cdot}26757969 \times 10^{3}$	$0{\cdot}69691719 \times 10^{-7}$
7	$0{\cdot}45724621 \times 10^{-3}$	$0{\cdot}45724737 \times 10^{-3}$	16	$0{\cdot}74870726 \times 10^{6}$	$0{\cdot}23230573 \times 10^{-7}$
8	$0{\cdot}15240245 \times 10^{-3}$	$0{\cdot}15241579 \times 10^{-3}$			
9	$0{\cdot}50638806 \times 10^{-4}$	$0{\cdot}50805263 \times 10^{-4}$			
10	$0{\cdot}14701873 \times 10^{-4}$	$0{\cdot}16935088 \times 10^{-4}$	20	$0{\cdot}98606210 \times 10^{49}$	$0{\cdot}28679720 \times 10^{-9}$

TABLE 2.3b

n	$y_n^{(1)}$	$y_n^{(2)}$	$y_n^{(3)}$	$y_n^{(4)}$	$y_n^{(5)}$
1	0·319444	0·332448	0·333272	0·333329	0·333333
2	0·104032	0·110620	0·111076	0·111108	0·111111
3	$0{\cdot}343138 \times 10^{-1}$	$0{\cdot}368427 \times 10^{-1}$	$0{\cdot}370223 \times 10^{-1}$	$0{\cdot}370361 \times 10^{-1}$	$0{\cdot}370370 \times 10^{-1}$
4	$0{\cdot}114059 \times 10^{-1}$	$0{\cdot}122791 \times 10^{-1}$	$0{\cdot}123411 \times 10^{-1}$	$0{\cdot}123454 \times 10^{-1}$	$0{\cdot}123457 \times 10^{-1}$
5	$0{\cdot}380789 \times 10^{-2}$	$0{\cdot}409413 \times 10^{-2}$	$0{\cdot}411380 \times 10^{-2}$	$0{\cdot}411513 \times 10^{-2}$	$0{\cdot}411522 \times 10^{-2}$
6	$0{\cdot}127409 \times 10^{-2}$	$0{\cdot}136534 \times 10^{-2}$	$0{\cdot}137133 \times 10^{-2}$	$0{\cdot}137172 \times 10^{-2}$	$0{\cdot}137174 \times 10^{-2}$
7	$0{\cdot}426692 \times 10^{-3}$	$0{\cdot}455353 \times 10^{-3}$	$0{\cdot}457130 \times 10^{-3}$	$0{\cdot}457240 \times 10^{-3}$	$0{\cdot}457247 \times 10^{-3}$
8	$0{\cdot}142927 \times 10^{-3}$	$0{\cdot}151863 \times 10^{-3}$	$0{\cdot}152383 \times 10^{-3}$	$0{\cdot}152414 \times 10^{-3}$	$0{\cdot}152416 \times 10^{-3}$
9	$0{\cdot}478669 \times 10^{-4}$	$0{\cdot}506447 \times 10^{-4}$	$0{\cdot}507963 \times 10^{-4}$	$0{\cdot}508048 \times 10^{-4}$	$0{\cdot}508052 \times 10^{-4}$
10	$0{\cdot}160249 \times 10^{-4}$	$0{\cdot}168885 \times 10^{-4}$	$0{\cdot}169327 \times 10^{-4}$	$0{\cdot}169350 \times 10^{-4}$	$0{\cdot}169351 \times 10^{-4}$
11	$0{\cdot}536259 \times 10^{-5}$	$0{\cdot}563149 \times 10^{-5}$	$0{\cdot}564436 \times 10^{-5}$	$0{\cdot}564500 \times 10^{-5}$	$0{\cdot}564503 \times 10^{-5}$
12	$0{\cdot}179379 \times 10^{-5}$	$0{\cdot}187773 \times 10^{-5}$	$0{\cdot}188149 \times 10^{-5}$	$0{\cdot}188167 \times 10^{-5}$	$0{\cdot}188168 \times 10^{-5}$
13	$0{\cdot}599790 \times 10^{-6}$	$0{\cdot}626066 \times 10^{-6}$	$0{\cdot}627175 \times 10^{-6}$	$0{\cdot}627223 \times 10^{-6}$	
14	$0{\cdot}200483 \times 10^{-6}$	$0{\cdot}208733 \times 10^{-6}$	$0{\cdot}209061 \times 10^{-6}$		
15	$0{\cdot}669920 \times 10^{-7}$	$0{\cdot}695902 \times 10^{-7}$			
16	$0{\cdot}223797 \times 10^{-7}$				

2.5 High order recurrence relations

In this section, we consider the extension of some of the basic techniques developed in the previous sections to the numerical solution of recurrence relations of order greater than two. The general "state of the art" here is not as satisfactory as in the second order case, due mainly to the fact that Olver's algorithm has not been generally extended to equations of order greater than two. At the present time, however, the theory of linear high order recurrence relations is under intensive development and one can confidently predict the emergence of improved algorithms in the near future. We begin with an examination of the linear case and outline some of the attempts that have been made to extend the Olver approach. In the design of algorithms suitable for computing non-dominant solutions of linear recurrence relations the usual approach is to rewrite the original initial value problem as an equivalent boundary value problem with the number of initial conditions which are abandoned being dependent on the particular solution of the recurrence relation which is required. It seems difficult in general to derive useful sufficient conditions under which the required solution of our original initial value problem is approximated arbitrarily closely by that of the re-posed boundary value problem. If, however, we are able to choose the basis solutions of our recurrence relation so that $y_{i,n}$ dominates $y_{i+1,n}$ for $i = 1, 2, \ldots, k - 1$, an example of this is if all the solutions of our recurrence relation are of the form $\alpha_i b_i^n$ where the b_i are real and distinct in modulus and the α_i are constants, then we are able to state that the solution of our re-posed problem is a good approximation to that of the original one providing we choose our boundary conditions correctly (Oliver, 1968a). In the particular case where the required solution is second in the order of dominance, it is possible to extend Olver's algorithm directly to find the required solution and also the optimal value of N. The appropriate procedure is to abandon a single initial condition and to reset it as $y_N = 0$ for some large N. By following through exactly the same analysis as for the second order case an expression for the truncation error of the scheme may be obtained. This may be used to estimate the optimal value of N before any back substitution has been performed. In cases where the required solution is not the second in the order of dominance, we may still replace the original problem by an equivalent boundary value problem, but it is not possible to apply Olver's technique for estimating the optimal value of N. We are thus usually required to guess a value for N initially. A practical procedure which we could use would be to compute solution sequences for increasing integer values of N with the procedure being stopped, and the final solution sequence being accepted, when two successive sequences are in some sense sufficiently close. As mentioned previously, however, this procedure may be very wasteful as regards computing time, especially if the initial guessed value of N is a long way from the optimal value. A procedure which partly overcomes this difficulty is one which has been suggested by Oliver (1968b). Oliver's algorithm requires an estimate of the optimal value of N to be available initially, with all the inherent difficulties which are present if this estimate is a long way from the true optimal value. A sequence of better approximations to this optimal value is then built up at a relatively cheap computational cost. Oliver's algorithm in effect requires a solution sequence $y_n^{[N]}$ to be computed for each trial value of N, and for all $n \in [N_0, N]$ where N_0 is the point at which the first initial condition is specified. The solution sequence $y_n^{[\bar{N}]}$ is accepted as a final approximation if

$$\left|y_n^{[\bar{N}-1]} - y_n^{[\bar{N}]}\right| < \varepsilon,$$

where ε is some appropriate error tolerance. Although this procedure of accepting an iterate as the final solution if it and the previous iterate differ by less than some prescribed amount is an accepted computational procedure it does not guarantee that the solution obtained is in fact the required solution. In an attempt to establish an algorithm which is based on a firmer theoretical footing, and which does not require a value of N to be guessed in advance, Cash (1978d) has proposed an algorithm which, like Olver's algorithm, seeks to obtain an estimate of the truncation error before any back substitution has been performed. The main ideas behind the Cash algorithm are as follows. Consider the solution of the mth order linear recurrence relation

$$\sum_{j=0}^{m} \alpha_j(n) y_{n+j} = f(n)$$

with the boundary conditions

$$y_j = k_j, \qquad j = 1(1)q,$$

$$y_{N+j} = 0, \qquad j = 0(1)m-1-q.$$

The matrix set of equations defining the required solution of this problem is

$$\alpha_q y_{q+1}^{[N]} + \alpha_{q+1} y_{q+2}^{[N]} + \cdots + \alpha_{m-1} y_m^{[N]} + \alpha_m y_{m+1}^{[N]} = f(1) - \alpha_0 y_1 - \cdots - \alpha_{q-1} y_q$$

$$\alpha_{q-1} y_{q+1}^{[N]} + \alpha_q y_{q+2}^{[N]} + \cdots + \alpha_{m-1} y_{m+1}^{[N]} + \alpha_m y_{m+2}^{[N]} = f(2) - \alpha_0 y_2 - \cdots - \alpha_{q-2} y_q$$

$$\cdots \qquad \cdots \qquad \cdots$$

$$\alpha_0 y_{N-q-1}^{[N]} + \alpha_1 y_{N-q}^{[N]} + \cdots + \alpha_q y_{N-1}^{[N]} = f(N-q-1).$$

By using simple Gaussian elimination the sub-diagonal terms may be annihilated to yield an upper triangular system of equations of the form

$$\hat{a}_{q+1,q+1}y_{q+1}^{[N]} + \hat{a}_{q+1,q+2}y_{q+2}^{[N]} + \cdots + \hat{a}_{q+1,m+1}y_{m+1}^{[N]} = \hat{e}_{q+1},$$

$$\hat{a}_{q+2,q+2}y_{q+2}^{[N]} + \cdots + \hat{a}_{q+2,m+1}y_{m+1}^{[N]} + \hat{a}_{q+2,m+2}y_{m+2}^{[N]} = \hat{e}_{q+2},$$

$$\cdots\cdots\cdots\cdots\cdots$$

$$\hat{a}_{N-1,N-1}y_{N-1}^{[N]} = \hat{e}_{N-1}.$$

If now we define a sequence z_n^r as the solution of the adjoint equation

$$\hat{a}_{n,n+m-q}z_n^r + \hat{a}_{n+1,n+m-q}z_{n+1}^r + \cdots + \hat{a}_{n+m-q,n+m-q}z_{n+m-q}^r = 0$$

with initial conditions

$$z_{r-m+q+1}^r = z_{r-m+q+2}^r = \cdots = z_{r-1}^r = 0, \qquad z_r^r = 1,$$

it may be shown that

$$y_r^{[N+1]} - y_r^{[N]} = \hat{e}_N z_N^r / \hat{a}_{r,r}.$$

This last expression may be used to estimate the optimal value of N in advance. For a more detailed explanation of this algorithm and for several practical applications the reader is referred to Cash (1978d). Although this algorithm is not always as satisfactory as Olver's algorithm, if the solution is required over some complete range $0 \le n \le L$, it is satisfactory in cases where the solution is required at one particular point to a given degree of precision. It will be seen later that this is an important practical case in the solution of linear stiff systems of O.D.E.s, since we shall often need to estimate the local truncation error at each steppoint before deciding which steplength to use for the next integration step.

Before considering the extension of some of the algorithms developed in the previous section to higher order non-linear equations we first of all consider a somewhat different approach to the numerical solution of linear recurrence relations. To do this we shall need to re-examine some of the iterative schemes developed earlier and we start off by re-considering the equation

$$a_2(n)y_{n+2} + a_1(n)y_{n+1} + a_0(n)y_n = g(n),$$

which was discussed in Section 2.1.

We shall first need to examine the origins of the scheme defined by (2.16), (2.17) and explain why we made the particular choice for $b(n)$ defined by (2.17). Suppose now that we wish to compute a solution of (2.1) using an iterative scheme of the form

$$a_2(n)y_{n+2}^{(t-1)} + b(n)y_{n+1}^{(t)} + a_0(n)y_n^{(t)} = g(n) + [b(n) - a_1(n)]y_{n+1}^{(t-1)}, \quad (2.28)$$

where at this stage $b(n)$ is an arbitrary sequence depending on the integer variable n. Putting $t = 1$, $n = 0$ in (2.28) and taking as initial conditions $y_0^{(t)} = y_0$ for all $t > 0$ and $\{y_n^{(0)}\} = 0$ for all $n \geq 1$ we have

$$y_1^{(1)} = \frac{-a_0(0)}{b(0)}y_0 + \frac{g(0)}{b(0)}.$$

Suppose now that we define an error $\eta_{t,n}$ as

$$\eta_{t,n} = y_n - y_n^{(t)}. \quad (2.29)$$

Subtraction of (2.28) from the equation

$$a_2(n)y_{n+2} + b(n)y_{n+1} + a_0(n)y_n = g(n) + [b(n) - a_1(n)]y_{n+1}, \quad (2.30)$$

which is just a trivial rearrangement of (2.1), and use of (2.29) gives

$$a_2(0)\eta_{0,2} + b(0)\eta_{1,1} = [b(0) - a_1(0)]\eta_{0,1}. \quad (2.31)$$

In general there would not seem to be any advantage to be gained in using one particular value of $b(0)$ rather than another, unless we have some additional information regarding a possible relationship between $\eta_{0,1}$ and $\eta_{0,2}$. Thus in order to simplify (2.31) we choose

$$b(0) = a_1(0)$$

so that (2.31) becomes

$$\eta_{1,1} = -\frac{a_2(0)\eta_{0,2}}{b(0)}. \quad (2.32)$$

Subtracting (2.28) with $t = 1$, $n = 1$ from (2.30) we have

$$a_2(1)\eta_{0,3} + b(1)\eta_{1,2} + a_0(1)\eta_{1,1} = [b(1) - a_1(1)]\eta_{0,2}.$$

Using (2.32) this relation becomes

$$a_2(1)\eta_{0,3} + b(1)\eta_{1,2} = \left[b(1) - a_1(1) + \frac{a_2(0)a_0(1)}{b(0)}\right]\eta_{0,2}.$$

If we now choose

$$b(1) = a_1(1) - \frac{a_2(0)a_0(1)}{b(0)}$$

we have

$$\eta_{1,2} = \frac{-a_2(1)\eta_{0,3}}{b(1)}.$$

Continuing in this way we find that if we choose

$$\left.\begin{aligned} b(n) &= a_1(n) - \frac{a_0(n)a_2(n-1)}{b(n-1)} \\ \text{we have} & \\ \eta_{1,n} &= \frac{-a_2(n-1)}{b(n-1)}\eta_{0,n+1}. \end{aligned}\right\} \tag{2.33}$$

Similarly for general t it can be shown that

$$\eta_{t,n} = (-1)^t \eta_{0,n+t} \prod_{i=0}^{t-1} \frac{a_2(n-1+i)}{b(n-1+i)}. \tag{2.34}$$

In this way we see how, by performing an error analysis on our iteration scheme, we may derive a scheme of the general form (2.28) with a particularly simple expression for the truncation error at any stage. We note that, with our given initial conditions $y_n^{(0)} \equiv 0$, we have $\eta_{0,n} \equiv y_n$. Hence if, as a result of some extra information, it is known that $|y_n| \leq M$ say for all n, a sufficient condition for the iteration scheme (2.28) to converge to any given degree of precision at the point n is

$$\lim_{t\to\infty} \prod_{i=0}^{t} \frac{a_2(n-1+i)}{b(n-1+i)} = 0.$$

Since this expression depends only on the coefficients of the recurrence relation and not on the solution itself, we may determine whether or not our iteration scheme will converge before any iterations have actually been performed. This scheme may be extended to higher order equations in a straightforward way, but this will not be done in this section since our aim is merely to point out how iteration schemes of this form may be established, by first of all carrying out an error analysis. The hope is, of course, of finding schemes which are of more practical use.

It is of interest to note in passing that in the constant coefficient, homogeneous case we may extend our error analysis even further and by so doing derive a scheme which has marked similarities with Newton's method for the determination of zeros of polynomials. This scheme will be discussed briefly since it introduces a new concept whereby the arbitrary sequence b

brought into the iteration scheme is no longer a function of n as it was in (2.28) but is a function of the iteration parameter t. To introduce our method of approach we consider the recurrence relation

$$y_{n+2} + by_{n+1} + cy_n = 0 \tag{2.35}$$

and attempt to generate a solution using the iterative scheme

$$y_{n+2}^{(t-1)} + b(t)y_{n+1}^{(t)} + cy_n^{(t)} = (b(t) - b)y_{n+1}^{(t-1)}. \tag{2.36}$$

We shall assume that associated with this scheme we are given just one initial condition of the form $y_0 = k_0$ and we shall take $y_0^{(t)} = k_0$ for all $t \geq 0$. We shall further assume that we wish to calculate a solution of the form $y_n = k_0\beta^n$, i.e. a solution which is directly proportional to just one of the basis solutions. The fact that our arbitrary sequence b is a function of t and not of n reflects the fact that the coefficients of our recurrence relation are constant. Thus, once y_1 is known we know y_n for all n. Using the initial approximation $y_2^{(0)} = 0$, putting $b(1) = b$ and subtracting (2.36) from (2.35), we have for $n = 0, t = 1$

$$y_1^{(1)} = -\frac{c}{b(1)}y_0$$

and so

$$\eta_{1,1} = -\frac{\eta_{0,2}}{b(1)}.$$

Since we know that our required solution has the property that $y_2 = y_1^2$, we choose $y_2^{(1)} = [y_1^{(1)}]^2$ because this will give the correct value for y_2 once y_1 itself is correct. Now from (2.29) for $t = 2$, $n = 1$ we have

$$\begin{aligned}
\eta_{2,1} &= y_1 - y_1^{(2)}, \\
&= y_1 - \frac{1}{b(2)}\{(b(2) - b)y_1^{(1)} - y_2^{(1)} - cy_0\}, \qquad \text{from (2.36)}, \\
&= \frac{b(2)y_1 - (b(2) - b)[y_1 - \eta_{1,1}] + y_2 - \eta_{1,2} + cy_0}{b(2)}, \\
&= \frac{by_1 + (b(2) - b)\eta_{1,1} + y_2 - \eta_{1,2} + cy_0}{b(2)}, \\
&= \frac{(b(2) - b)\eta_{1,1} - \eta_{1,2}}{b(2)}, \qquad \text{from (2.35)}, \\
&= \frac{(b(2) - b)\eta_{1,1} - (y_2 - [y_1^{(1)}]^2)}{b(2)},
\end{aligned}$$

$$= \frac{(b(2) - b)\eta_{1,1} - 2(y_1^{(1)} + \eta_{1,1})\eta_{1,1} + \eta_{1,1}^2}{b(2)},$$

$$= \frac{[b(2) - b - 2y_1^{(1)}]\eta_{1,1} - \eta_{1,1}^2}{b(2)}.$$

Thus if we choose $b(2) = b + 2y_1^{(1)}$ we have

$$\eta_{2,1} = -\frac{\eta_{1,1}^2}{b(2)},$$

i.e. quadratic convergence. In general, if we choose

$$b(t) = b + 2y_1^{(t-1)}, \qquad y_2^{(t-1)} = [y_1^{(t-1)}]^2$$

we have

$$\eta_{t,1} = -\frac{\eta_{t-1,1}^2}{b(t)}.$$

This scheme may be regarded as a form of stationary iteration, since we do not compute the solution for increasing values of n but only calculate y_1 and y_2. This algorithm is in effect just another form of Newton iteration for the solution of the polynomial equation $x^2 + bx + c = 0$ and consequently we shall not consider it any further. The algorithm does, however, establish the interesting precedent that the sequence b may be allowed to depend on the iteration parameter t. An obvious extension of the two approaches considered in this section would be to develop an iteration scheme of the form (2.28) where the sequence b would be allowed to depend on both t and n. This is an approach which would seem to merit further investigation.

An important property of the iterative algorithms which have been developed in this chapter is that they may be extended to higher order linear recurrence relations. To illustrate this extension we consider

$$a_3(n)y_{n+3} + a_2(n)y_{n+2} + a_1(n)y_{n+1} + a_0(n)y_n = 0. \qquad (2.37a)$$

As was mentioned previously it is quite straightforward to extend the iteration scheme (2.28) to generate a non-dominant solution of (2.37a) in cases where direct methods of solution are unstable. Assuming that we are able to choose three basis solutions $y_{1,n}$, $y_{2,n}$ and $y_{3,n}$, which are such that

$$\lim_{n\to\infty} \frac{y_{2,n}}{y_{1,n}} = \lim_{n\to\infty} \frac{y_{3,n}}{y_{1,n}} = \lim_{n\to\infty} \frac{y_{3,n}}{y_{2,n}} = 0,$$

we could seek to generate a solution of (2.37a) which is directly proportional to

$y_{3,n}$ by using an iteration scheme of the form

$$a_3(n)y_{n+3}^{(t-1)} + a_2(n)y_{n+2}^{(t-1)} + b(n)y_{n+1}^{(t)} + a_0(n)y_n^{(t)} = [b(n) - a_1(n)]y_{n+1}^{(t-1)}. \quad (2.37b)$$

We could then attempt to generate a solution of (2.37a) depending on $y_{2,n}$ and $y_{3,n}$ only (i.e. not containing a component of the dominant solution $y_{1,n}$) using an iteration scheme of the form

$$a_3(n)y_{n+3}^{(t-1)} + b(n)y_{n+2}^{(t)} + a_1(n)y_{n+1}^{(t)} + a_0(n)y_n^{(t+1)} = [b(n) - a_2(n)]y_{n+2}^{(t-1)}, \quad (2.37c)$$

using appropriate choices for the sequence $b(n)$. If we consider first of all scheme (2.37c) we see that in order to be able to apply this scheme to the solution of (2.37a) we need two initial conditions of the form

$$y_0 = k_0, \qquad y_1 = k_1$$

to be specified. We then set $y_0^{(t)} = k_0, y_1^{(t)} = k_1$ for all integers $t \geq 0$. By performing an analysis of the truncation error of scheme (2.37c) we may derive an algorithm which is a natural extension to the third order case of the algorithm defined by (2.16), (2.17). An important point to note about scheme (2.37c) is that it involves the three iteration levels $t - 1, t, t + 1$. This is to be expected, however, since it corresponds to iterating along lines $n + t = k$ for $k = 3, 4, \ldots$ in a similar fashion to that defined in Fig. 2.1. As such it is a direct extension of the procedure adopted in the second order case. It is straightforward to show that if we make the choice

$$\left.\begin{aligned} &b(n) = a_2(n) - \frac{a_1(n)a_3(n-1)}{b(n-1)} + \frac{a_0(n)a_3(n-1)a_3(n-2)}{b(n-1)b(n-2)} \\ &\qquad \text{for all } n \geq 0 \\ &\text{where} \\ &b(-1) = \infty \end{aligned}\right\} \quad (2.38)$$

and we define an error $\eta_{t,n}$ as in (2.29) then

$$\begin{aligned} \eta_{t,n} &= -\frac{a_3(n-2)}{b(n-2)}\eta_{t-1,n+1}, \\ &= (-1)^t \eta_{0,n+t} \prod_{i=1}^{t} \frac{a_3(n-3+i)}{b(n-3+i)}. \end{aligned}$$

In particular if we make the initial approximations $y_n^{(0)} \equiv 0$ for all $n \geq 1$ and we know from some additional source that $|y_n| \leq M$ for all n, then it is possible to obtain convergence to an absolute accuracy ε if there exists a value t such that

$$\prod_{i=1}^{t} \left| \frac{a_3(n-3+i)}{b(n-3+i)} \right| < \varepsilon/M.$$

Considering now scheme (2.37b) we see that in order to use this scheme we need one initial condition of the form

$$y_0^{(t)} = k_0 \quad \text{for all integers} \quad t \geq 0.$$

By carrying out an error analysis of this scheme we may show that if we choose

$$\left.\begin{aligned} &b(-1) = \infty, \\ &b(n) = a_1(n) - \frac{a_0(n)a_2(n-1)}{b(n-1)} + \frac{a_0(n)a_0(n-1)a_3(n-2)}{b(n-1)b(n-2)} \\ &\qquad \text{for all } n \geq 0 \end{aligned}\right\} \tag{2.39}$$

then

$$\eta_{t,n+1} = -\frac{a_3(n)\eta_{t-1,n+2} - \left(a_2(n) - \dfrac{a_0(n)a_3(n-1)}{b(n-1)}\right)\eta_{t-1,n+1}}{b(n)}. \tag{2.40}$$

We may now state the following theorem regarding the convergence of scheme (2.37b) for any choice of initial approximations $\{y_n^{(0)}\}$; see also Cash (1977a).

Theorem 2.5
Consider the iterative solution of (2.37a) *using* (2.37b) *where the sequence* $b(n)$ *is chosen to satisfy* (2.39). *Then if*

$$\left| \frac{a_3(n)}{b(n)} \right| + \left| \frac{a_2(n) - \dfrac{a_0(n)a_3(n-1)}{b(n-1)}}{b(n)} \right| \leq P < 1 \quad \text{for all } n$$

it follows that (2.37b) *converges to a solution of* (2.37a) *for any choice of the initial sequence* $\{y_n^{(0)}\}$, $n \geq 1$, *and furthermore*

$$\max_n |\eta_{t,n}| \leq P^t \max_n |\eta_{0,n}|.$$

The proof of this is an immediate extension of Lemma 2.2 and so will not be given here. In order to illustrate the two algorithms (2.37b) and (2.37c) we consider a particular example.

Example 2.4
Consider the linear recurrence relation

$$y_{n+3} - 111y_{n+2} + 1110y_{n+1} - 1000y_n = 0.$$

(a) Generate the minimal solution in the range $1 \leq n \leq 6$ using the initial condition $y_0 = 1$.

(b) Generate a solution not containing a component of the dominant solution in the range $2 \leq n \leq 6$ using the initial conditions $y_0 = 3/2$, $y_1 = 6$.

The results obtained for problem (a) using initial approximations $\{y_n^{(0)}\} \equiv 0$ are given in Table 2.4a. As can be seen convergence to the required solution $y_n \equiv 1$ is obtained in about six iterations with the rate of convergence becoming progressively slower as n increases. The results obtained for problem (b) using the same initial conditions are given in Table 2.4b. Convergence to the required solution $y_n = 1 + \frac{1}{2} \times 10^n$ is again obtained in six iterations. By carrying out the iteration schemes for larger values of n it is found that both schemes remain stable for all n.

TABLE 2.4a

n	$y_n^{(1)}$	$y_n^{(2)}$	$y_n^{(3)}$	$y_n^{(4)}$	$y_n^{(5)}$	$y_n^{(6)}$
1	0·900901	0·990099	0·999089	0·999926	0·999995	1·000000
2	0·891981	0·989914	0·999182	0·999947	0·999998	1·000000
3	0·891098	0·990696	0·999371	0·999976	1·000001	1·000000
4	0·891010	0·991576	0·999570	1·000004	1·000005	1·000001
5	0·891000	0·992466	0·999770	1·000032	1·000007	1·000001
6	0·891000	0·993357	0·999971	1·000086	1·000010	1·000001

It is clear that this iterative approach may be extended to deal with higher order linear recurrence relations but since this extension is straightforward we shall not give it explicitly in this chapter.

2.6 A quadratic factor approach

As an alternative method of approach to the numerical solution of linear recurrence relations we could seek to extract from (2.37a) a "quadratic factor" by defining an iteration scheme on the coefficients of our recurrence relation rather than on the solution y_n. It will be seen that this approach is rather in the spirit of Bairstow's method for the constant coefficient case. The approach

TABLE 2.4b

n	$y_n^{(1)}$	$y_n^{(2)}$	$y_n^{(3)}$	$y_n^{(4)}$	$y_n^{(5)}$	$y_n^{(6)}$
2	46·4865	50·5539	50·9554	50·9955	50·9996	51·0000
3	451·485	496·054	500·505	500·951	500·995	501·000
4	4501·48	4951·05	4996·01	5000·50	5000·95	5001·00
5	45001·5	49501·1	49951·0	49996·0	50000·5	50001·0
6	450001	495001	499501	499951	499996	500001

would prove valuable if the conditions associated with (2.37a) were two-point boundary conditions in cases where only two conditions are specified but we are able to pick out the required solution using some additional information regarding its behaviour. This approach will prove useful in the solution of two-point boundary value problems for ordinary differential equations. For the moment we assume that the $a_i(n)$ and y_n are scalar functions of the non-negative integer variable n and also that it is possible to choose a basis $(y_{1,n}, y_{2,n}, y_{3,n})$ for (2.37a) having the property that

$$\lim_{n\to\infty} \frac{y_{2,n}}{y_{1,n}} = \lim_{n\to\infty} \frac{y_{3,n}}{y_{1,n}} = 0. \tag{2.41}$$

Note that no assumptions have been made regarding the relative behaviour of $y_{2,n}$ and $y_{3,n}$. Any solution y_n of (2.37a) may be written in the form

$$y_n = \sum_{i=1}^{3} c_i y_{i,n} \tag{2.42}$$

and it will be assumed that the required solution takes this form with $c_1 = 0$. Suppose now that we wish to extract from (2.37a) a quadratic factor of the form

$$y_{n+2} = a(n)y_{n+1} + b(n)y_n. \tag{2.43}$$

Substituting (2.43) into (2.37a) we obtain the relation

$$\{a_3(n)[a(n+1)a(n) + b(n+1)] + a_2(n)a(n) + a_1(n)\}y_{n+1}$$
$$+ \{a_3(n)a(n+1)b(n) + a_2(n)b(n) + a_0(n)\}y_n = 0.$$

Since this first order, homogeneous equation is satisfied by two linearly independent solutions of (2.43), it follows that

$$\left.\begin{aligned} a_3(n)[a(n+1)a(n) + b(n+1)] + a_2(n)a(n) + a_1(n) &= 0, \\ a_3(n)a(n+1)b(n) + a_2(n)b(n) + a_0(n) &= 0. \end{aligned}\right\} \tag{2.44}$$

In the constant coefficient case we have $a(n+1) = a(n) = a$ and $b(n+1) = b(n) = b$ and so (2.44) defines two non-linear equations in the two unknowns a and b. We may generate a solution of these equations using Newton's method and this whole scheme is, of course, precisely Bairstow's method. In the case of variable coefficients, however, we have more unknowns than equations, because there clearly cannot be a general relationship between $a(n+1)$ and $a(n)$ or $b(n+1)$ and $b(n)$. Thus we have to consider a different method for the solution of (2.44) from that used in the constant coefficient case. In this section we shall consider two particular methods for the solution of

(2.44), one of which arises as a result of the particular form of the equations which occur in the intended practical applications, and the other being an attempt to extend Olver's approach to deal with quadratic factors.

The first method of solution arises as a result of noticing that in one of the applications for which we shall use this approach, namely in the numerical solution of two-point boundary value problems for ordinary differential equations, the coefficient $a_3(n)$ is usually very small and $a_2(n)$ is non-zero. As a result a good first approximation to the required solutions $a(n)$ and $b(n)$ is usually given by

$$a(n) = \frac{-a_1(n)}{a_2(n)},$$

$$b(n) = \frac{-a_0(n)}{a_2(n)},$$

which arise as a result of taking $a_3(n) \equiv 0$ in (2.44). This in turn suggests the use of the iteration scheme

$$\left.\begin{aligned} &a_3(n)\{a^{(t-1)}(n+1)a^{(t)}(n) + b^{(t-1)}(n+1)\} + a_2(n)a^{(t)}(n) + a_1(n) = 0, \\ &a_3(n)a^{(t-1)}(n+1)b^{(t)}(n) + a_2(n)b^{(t)}(n) + a_0(n) = 0. \end{aligned}\right\} \quad (2.45)$$

Useful sufficient conditions for the convergence of the iteration scheme (2.45) have proved to be rather difficult to derive since it would seem that any criterion for convergence must depend on $a(n)$ and $b(n)$ and neither of these two quantities is known initially. Practical experience has shown, however, that, in cases where for all n $a_3(n)$ is very small compared with $a_2(n)$, the scheme (2.45) usually does converge fairly rapidly to the required solution. We shall not attempt to analyse scheme (2.45), although there is a need for some theoretical results regarding the convergence properties of this scheme, but shall instead present some numerical evidence at a later stage.

As a second method for the solution of (2.44) we consider an extension of Olver's technique described in Section 2.1. Suppose that for some large integer N we set $a(N+1) = b(N+1) = 0$, where N is at this stage arbitrary. Substitution into (2.44) for $n = N$ gives

$$a_2(N)a(N) + a_1(N) = 0,$$

$$a_2(N)b(N) + a_0(N) = 0,$$

and so

$$a(N) = -\frac{a_1(N)}{a_2(N)}, \qquad b(N) = -\frac{a_0(N)}{a_2(N)}.$$

Substituting into (2.44) for $n = N - 1$ we have

$$a_3(N-1)[a(N)a(N-1) + b(N)] + a_2(N-1)a(N-1) + a_1(N-1) = 0,$$
$$a_3(N-1)a(N)b(N-1) + a_2(N-1)b(N-1) + a_0(N-1) = 0, \qquad (2.46)$$

and since $a(N)$ and $b(N)$ have already been calculated we may solve for $a(N-1)$ and $b(N-1)$ immediately. Note that it is not necessary to solve simultaneous equations for $a(N-1)$ and $b(N-1)$ since we may solve the second of (2.46) for $b(N-1)$ immediately and we may then solve the first of (2.46) for $a(N-1)$ and so in practice these equations are not simultaneous but are uncoupled. Thus we may write the general solution of our equations (2.44) as

$$\left.\begin{aligned} a(n) &= -\left\{\frac{a_1(n) + a_3(n)b(n+1)}{a_2(n) + a_3(n)a(n+1)}\right\}, \\ b(n) &= -\left\{\frac{a_0(n)}{a_3(n)a(n+1) + a_2(n)}\right\}. \end{aligned}\right\} \qquad (2.47)$$

Unfortunately the convergence properties of this scheme once again seem to be rather difficult to analyse mainly due to the fact that equations (2.47) are non-linear.

The analysis given by Olver for linear second order recurrence relations is particularly effective mainly on account of the fact that he is able to derive a linear relationship between the tth iterates, $y_n^{[t]}$ and $y_{n+1}^{[t]}$, at any two successive points in the region of interest (see equation (2.8)). This cannot easily be done for higher order recurrence relations, however, and as a result there is still a need for a more firm theoretical basis for our algorithm. It is hoped, however, that the numerical results which we shall present are sufficiently convincing to indicate that this line of approach is worth pursuing. Suppose now that we wish to calculate the required solution in the range $0 \le n \le L$ correct to D decimal places for given values of the integers L and D. We denote the solutions obtained from (2.44) as a result of setting $a(N+1) = b(N+1) = 0$ by $a^{[N]}(n)$ and $b^{[N]}(n)$. What we would ideally like to be able to do is to estimate the first value of N for which

$$|a^{[N]}(n) - a^{[N+1]}(n)| < \tfrac{1}{2} \times 10^{-D},$$
$$|b^{[N]}(n) - b^{[N+1]}(n)| < \tfrac{1}{2} \times 10^{-D},$$

for all n in the region of interest, before any back substitution has been performed. As already mentioned, however, this turns out to be difficult because of the complicated nature of the expression for the truncation error of our scheme. In view of this the procedure actually adopted in practice is the well-known one whereby we set $a(n+1) = b(n+1) = 0$ for some value $n = N$ and then we calculate the two sequences $a^{[N]}(n)$ and $b^{[N]}(n)$ for all $0 \leq n \leq N$. We now replace N by $N + 5$ and calculate another two sequences $a^{[N+5]}(n)$ and $b^{[N+5]}(n)$. If these two sequences for $a(n)$ and $b(n)$ differ by less than a prescribed amount, ε, in the sense that for all n

$$|a^{[N]}(n) - a^{[N+5]}(n)| < \varepsilon$$

and

$$|b^{[N]}(n) - b^{[N+5]}(n)| < \varepsilon$$

then we take the sequences $a^{[N+5]}(n), b^{[N+5]}(n)$ as our finally accepted sequence of approximations. If this is not the case we continue our procedure in an obvious way by starting at the point $N + 10$.

It is important to note that once we have obtained our final approximations to the sequences $a(n)$ and $b(n)$ we still cannot guarantee to find the required numerical solution to the specified degree of precision. Having obtained the quadratic factor (2.43) we may use the given two-point boundary conditions, which we assume to take the form

$$y(0) = y_0, \qquad y(\bar{N}) = y_{\bar{N}},$$

to rewrite the equations satisfied by the required solution of (2.37a) in the matrix form

$$\mathbf{Ay} = \mathbf{b},$$

where $\mathbf{y} = (y_1, y_2, \ldots, y_{\bar{N}-1})^T$, $\mathbf{b} = (-b(0)y_0, 0, \ldots, 0, -y_{\bar{N}})^T$ and where $\mathbf{A}$ is an appropriate tri-diagonal matrix depending on the two sequences $a(n)$ and $b(n)$ which were computed in the first part of the algorithm. A certain degree of imprecision in our final solution, $\mathbf{y}$, may arise as a result of the matrix $\mathbf{A}$ not being calculated exactly and possibly as a result of inexact boundary conditions. The effect on the solution of a system of linear algebraic equations due to perturbations in $\mathbf{A}$ and $\mathbf{b}$ is well known and is given for example in the following theorem, the proof of which may be found in Isaacson and Keller (1966), p. 37.

Theorem 2.6

Consider the solution of the linear system of algebraic equations $\mathbf{Ax} = \mathbf{b}$ *where the data* $\mathbf{A}$ *and* $\mathbf{b}$ *are perturbed by amounts* $\boldsymbol{\delta}\mathbf{A}$ *and* $\boldsymbol{\delta}\mathbf{b}$ *respectively giving rise to a perturbed solution* $\mathbf{x} + \boldsymbol{\delta}\mathbf{x}$ *which satisfies*

$$(\mathbf{A} + \boldsymbol{\delta}\mathbf{A})(\mathbf{x} + \boldsymbol{\delta}\mathbf{x}) = \mathbf{b} + \boldsymbol{\delta}\mathbf{b}.$$

Suppose now that $\mathbf{A}$ *is non-singular and that* $\boldsymbol{\delta}\mathbf{A}$ *is sufficiently small so as to satisfy the inequality*

$$\|\boldsymbol{\delta}\mathbf{A}\| < \frac{1}{\|\mathbf{A}^{-1}\|}.$$

Then

$$\frac{\|\boldsymbol{\delta}\mathbf{x}\|}{\|\mathbf{x}\|} \leq \frac{\mu}{1 - \mu\|\boldsymbol{\delta}\mathbf{A}\|/\|\mathbf{A}\|}\left[\frac{\|\boldsymbol{\delta}\mathbf{b}\|}{\|\mathbf{b}\|} + \frac{\|\boldsymbol{\delta}\mathbf{A}\|}{\|\mathbf{A}\|}\right],$$

where μ *is the condition number of* $\mathbf{A}$ *defined by*

$$\mu = \|\mathbf{A}^{-1}\| \cdot \|\mathbf{A}\|.$$

In view of this we are not able to guarantee that the relative error in $\mathbf{x}$ is of the same order of magnitude as that in $\mathbf{A}$ or $\mathbf{b}$. If, however, we are able by some other means to estimate the values of μ and $\|\mathbf{A}\|$ we may obtain some idea of the effect of perturbations in $\mathbf{A}$ and $\mathbf{b}$ on our final results. In any case, if our final results are required to an absolute accuracy of ε say, it is advisable to use rather more precision, say $\varepsilon/100$ in the absence of any further information, when computing $a(n)$ and $b(n)$. An approach which requires about twice as much computational effort but which gives us greater confidence in the results obtained is that of solving our system of algebraic equations in the case where the coefficient matrix depends on $a^{[N]}(n)$ and $b^{[N]}(n)$ as well as in the case where this matrix depends on $a^{[N+5]}(n)$ and $b^{[N+5]}(n)$. If the two sets of solutions obtained differ by less than the prescribed precision we can then be reasonably certain that we have the required solution. We now present a numerical example to illustrate these two approaches.

Example 2.5

The equation which we consider is based on the numerical solution of the first order ordinary differential equation

$$y' = -xy, \qquad y(0) = 1$$

using the finite difference equation

$$y_{n+2} + 18y_{n+1} - 9y_n - 10y_{n-1} = h\{9y'_{n+1} + 18y'_n + 3y'_{n-1}\}.$$

Applying this scheme to the given differential equation we obtain

$$y_{n+2} + (18 + 9hx_{n+1})y_{n+1} - (9 - 18hx_n)y_n - (10 - 3hx_{n-1})y_{n-1} = 0.$$

Suppose now that we attempt to extract a quadratic factor of this equation where for demonstration purposes we take $h = 0{\cdot}1$ and where the coefficients of the quadratic factor are required correct to six decimal places. In Table 2.5a we give the values obtained for these coefficients using iteration scheme (2.45) in the range $0 \leq n \leq 8$ and as can be seen convergence is obtained in six iterations. In Table 2.5b we give the results obtained using the algorithm based on (2.46) and (2.47). For the sake of brevity we shall not give the results obtained using different values of N but shall instead choose one particular value of N, $N = 14$ in this case, and compare the results obtained with those given in Table 2.5a. As can be seen there is a close agreement between the results obtained in the two tables.

Having obtained a quadratic factor of the form (2.43) we could now calculate an extra initial condition using a one-step starting procedure, such as Runge–Kutta for example, and then recur in the direction of increasing n to obtain the required solution. For the sake of our demonstration, however, we shall consider the artificial example where we assume that $y_{10} = \exp(-0{\cdot}5)$ is given and so we need to solve a system of algebraic equations for our required solution. It was found that, to six decimal places, the results obtained using the two methods were identical and in Table 2.5c we list these results as well as the true solution. As can be seen agreement is satisfactory.

Finally we mention the extension of our techniques to higher order non-linear recurrence relations. There is very little to add on this subject since the relevant extensions are immediate. Suppose for example we consider the equation

$$a_3(n)y_{n+3} + a_2(n)y_{n+2} + a_1(n)y_{n+1} + a_0(n)y_n = g(y_n, y_{n+1}, y_{n+2}, y_{n+3}).$$

Following the approach adopted in Section 2.4 we generate the required solution of this equation using two distinct iteration schemes—a primary and a secondary. As before we shall use a Gauss–Seidel scheme as the primary iteration and either quasi-Newton or direct iteration as our second scheme. We shall not consider these schemes any further at this stage but will discuss them more fully when they are applied to the solution of ordinary differential equations in Chapter 3.

TABLE 2.5a

n		1st Iteration	2nd Iteration	3rd Iteration	4th Iteration	5th Iteration	6th Iteration
0	$a(n)$	0·4851485	0·4435929	0·4452803	0·4451814	0·4451860	0·4451857
	$b(n)$	0·5500550	0·5361095	0·5372771	0·5372308	0·5372335	0·5372333
1	$a(n)$	0·4729064	0·4323687	0·4339748	0·4338812	0·4338854	0·4338852
	$b(n)$	0·5457033	0·5322788	0·5334052	0·5333616	0·5333641	0·5333640
2	$a(n)$	0·4607843	0·4212327	0·4227608	0·4226722	0·4226761	0·4226759
	$b(n)$	0·5413943	0·5284766	0·5295633	0·5295223	0·5295246	0·5295245
3	$a(n)$	0·4487805	0·4101838	0·4116373	0·4115535	0·4115571	0·4115570
	$b(n)$	0·5371274	0·5247025	0·5257512	0·5257126	0·5257148	0·5257147
4	$a(n)$	0·4368932	0·3992214	0·4006036	0·4005241	0·4005275	0·4005273
	$b(n)$	0·5329018	0·5209563	0·5219685	0·5219323	0·5219344	0·5219343
5	$a(n)$	0·4251208	0·3883447	0·3896584	0·3895832	0·3895863	0·3895862
	$b(n)$	0·5287171	0·5172379	0·5182151	0·5181810	0·5181830	0·5181829
6	$a(n)$	0·4134615	0·3775528	0·3788010	0·3787298	0·3787327	0·3787325
	$b(n)$	0·5245726	0·5135469	0·5144905	0·5144585	0·5144603	0·5144603
7	$a(n)$	0·4019139	0·3668449	0·3680305	0·3679630	0·3679656	0·3679655
	$b(n)$	0·5204678	0·5098832	0·5107945	0·5107645	0·5107662	0·5107662
8	$a(n)$	0·3904762	0·3562204	0·3573459	0·3572818	0·3572843	0·3572842
	$b(n)$	0·5164021	0·5062465	0·5071268	0·5070987	0·5071003	0·5071002

TABLE 2.5b

n			n		
0	$a(n)$	0·4451857	5	$a(n)$	0·3895862
	$b(n)$	0·5372334		$b(n)$	0·5181829
1	$a(n)$	0·4338852	6	$a(n)$	0·3787325
	$b(n)$	0·5333640		$b(n)$	0·5144603
2	$a(n)$	0·4226759	7	$a(n)$	0·3679655
	$b(n)$	0·5295245		$b(n)$	0·5107662
3	$a(n)$	0·4115570	8	$a(n)$	0·3572842
	$b(n)$	0·5257147		$b(n)$	0·5071002
4	$a(n)$	0·4005274			
	$b(n)$	0·5219343			

TABLE 2.5c

n	y_n	$y(x_n)$
1	0·9950125	0·9950125
2	0·9801987	0·9801987
3	0·9559976	0·9559975
4	0·9231164	0·9231163
5	0·8824970	0·8824969
6	0·8352703	0·8352702
7	0·7827046	0·7827045
8	0·7261491	0·7261490
9	0·6669768	0·6669768

2.7 Another class of iteration schemes

In the previous sections we considered in some detail certain classes of iteration schemes which were based, to a large extent, on a Gauss–Seidel type of approach. These classes of schemes were found to be generally applicable to the solution of linear recurrence relations. Some were also capable of direct extension to the non-linear case if a Gauss–Seidel scheme was used as the primary iteration with an alternative scheme being used as the secondary iteration. In practice the procedure which we adopt when using these iteration schemes is to apply them for increasing values of the iteration parameter until either convergence occurs or until a maximum allowable number of iterates have been performed (whichever is the sooner). As a result it is not known initially how many iterations we will need to carry out before convergence to the prescribed degree of precision is obtained. We might expect that there would be certain practical advantages if we could develop a class of iteration schemes which was guaranteed to converge to the prescribed degree of precision in a fixed number of primary iterations, especially if this number were 1. We would then hope that the computational effort involved with these schemes would be relatively small. In this section we develop one such class of iteration schemes which is specially designed for the solution of finite difference equations arising from the application of linear multistep methods to the solution of first order ordinary differential equations.

Suppose then that we wish to generate a solution of the kth order recurrence relation

$$\mathbf{R}(\mathbf{y}_n, \mathbf{y}_{n+1}, \ldots, \mathbf{y}_{n+k}) = 0, \tag{2.48}$$

where we know from some external source that the required solution of our problem may be approximated arbitrarily closely by the solution of an equivalent boundary value problem where i $(i < k)$ of the original initial conditions are retained whilst the other $k - i$ conditions are at the far end of the region of interest. We could attempt to generate the required solution of (2.48) using an iterative scheme of the form

$$\mathbf{F}(\mathbf{y}_n^{(t)}, \mathbf{y}_{n+1}^{(t)}, \ldots, \mathbf{y}_{n+i}^{(t)}) = \mathbf{G}(\mathbf{y}_n^{(t-1)}, \mathbf{y}_{n+1}^{(t-1)}, \ldots, \mathbf{y}_{n+k}^{(t-1)}). \tag{2.49}$$

In (2.49) the vector-valued functions **F** and **G** are at this stage almost arbitrary with the only restriction on **F** being that it must be such that we are able to solve for a unique value $\mathbf{y}_{n+i}^{(t)}$ for all integers i, t and the only restriction on **G** being that it must be chosen so that

$$\mathbf{F}(\mathbf{y}_n, \mathbf{y}_{n+1}, \ldots, \mathbf{y}_{n+i}) = \mathbf{G}(\mathbf{y}_n, \mathbf{y}_{n+1}, \ldots, \mathbf{y}_{n+k}).$$

The first condition guarantees the solvability of our problem. The second condition guarantees that the required solution of (2.48) is a solution of our iteration scheme. Suppose now that we are given i initial conditions of the form $\mathbf{y}_m = \mathbf{k}_m$, $m = 0(1)i - 1$, and that a sequence of approximations $\{\mathbf{y}_n^{(0)}\}$ is available for all $n \geq i$. Then on using the initial conditions $\mathbf{y}_m^{(t)} = \mathbf{k}_m$ for $0 \leq m \leq i - 1$ and all integers $t \geq 0$, we may use scheme (2.49) with $t = 1, 2, \ldots$ in succession to calculate sequences of approximations $\{\mathbf{y}_n^{(1)}\}, \{\mathbf{y}_n^{(2)}\} \ldots$ for all n in the range of interest. In order that this iteration scheme should be of any practical use for the solution of our problem we need to impose the additional requirement that

$$\lim_{t \to \infty} \mathbf{y}_n^{(t)} = \mathbf{y}_n \quad \text{for all } n = i, i + 1, \ldots$$

For the sake of efficiency we would also like this convergence to take place as rapidly as possible. It can be seen from the very general form of (2.49) that the possible choice of iteration schemes is still very wide at this stage, although in practice it is not easy to find rapidly convergent ones which do not require the initial approximations to be particularly accurate.

Our aim in this section is to develop a class of iteration schemes specially designed to generate a solution of a small class of recurrence relations of the form

$$\sum_{j=0}^{k} \alpha_j \mathbf{y}_{n+j} = h \sum_{j=0}^{k} \beta_j \mathbf{f}_{n+j}. \tag{2.50}$$

As a result we hope to obtain computational efficiency at the cost of more limited applicability. As will be shown in Chapter 3 the schemes which we develop do have important practical applications in connection with the numerical solution of stiff systems of first order ordinary differential equations. The particular class of schemes which we consider takes the general form

$$\sum_{j=0}^{i} \hat{\alpha}_j \mathbf{y}_{n+j}^{(t)} - h \sum_{j=0}^{i} \beta_j \mathbf{f}(\mathbf{y}_{n+j}^{(t)}) = \sum_{j=i-1}^{k-1} \tau_j \delta^2 \mathbf{y}_{n+j}^{(t-1)} + h \sum_{j=i+1}^{k} \beta_j \mathbf{f}(\mathbf{y}_{n+j}^{(t-1)}), \tag{2.51}$$

where the $\hat{\alpha}_j$'s and τ_j are constants, $\delta^2 \mathbf{y}_n = \mathbf{y}_{n+1} - 2\mathbf{y}_n + \mathbf{y}_{n-1}$ and where the integer i lies in the range $(1, k-1)$. This particular choice for our iteration scheme is rather hard to justify except for the fact that in practice it is found to satisfy our requirements. Another reason for this particular choice is that it is an attempt to extend an algorithm proposed by Miller (1966) for the solution of non-stiff problems and Miller's scheme is in fact a special case of (2.51). As mentioned previously, however, the possible choices for the functions **F** and **G** appearing in (2.49) are very large and it may well be the case that other choices would lead to a more efficient iteration scheme than that given by (2.51). It can be seen from (2.51) that we have in effect replaced our original initial value problem by a boundary value problem where only i of the original k initial conditions are used. If we now consider the case where (2.51) is to be used to generate the solution of a stiff system of first order ordinary differential equations, it is clear that there would be considerable practical advantages in letting i take the value 1 (or perhaps 2) since in this case the scheme would be self-starting and changes in the steplength of integration would be easy to perform. In Chapter 3 it will be shown that it is possible to develop efficient A-stable schemes, with $i = 1$ or 2, which are of use for the numerical integration of stiff systems of equations and which have order greater than two.

An examination of our iteration scheme reveals that if the coefficients $\hat{\alpha}_j$ and

τ_j appearing in (2.51) are chosen so as to satisfy the relations

$$\hat{\alpha}_j = \alpha_j, \quad 0 \leq j \leq i-3,$$

$$\tau_r = -\alpha_{r+1} + 2\tau_{r+1} - \tau_{r+2} + \hat{\alpha}_{r+1},$$

$$i+1 \leq r \leq k-1 \quad \text{with } \tau_k = \tau_{k+1} = 0,$$

$$\hat{\alpha}_{i-1} = \alpha_{i-1} + \tau_i - 2\tau_{i-1},$$

$$\hat{\alpha}_{i-2} = \alpha_{i-2} + \tau_{i-1},$$

then if the two-dimensional sequence $\mathbf{y}_n^{(t)}$ generated by scheme (2.51) converges to some final sequence $\mathbf{v}(n)$ in the sense that

$$\lim_{t\to\infty} \|\mathbf{y}_n^{(t)} - \mathbf{v}(n)\| = 0, \qquad n = i, i+1, \ldots,$$

then $\mathbf{v}(n)$ will also satisfy (2.50), since on convergence (2.51) reduces to (2.50) identically. As mentioned previously we are especially interested in developing iteration schemes of the form (2.51) which, when applied to the numerical solution of stiff systems, are not used in this particular way but which converge to the required solution of our system of differential equations in a fixed number of primary iterations. Since in this case we are not iterating to convergence, we shall not be solving our linear multistep method exactly. Thus the schemes developed will have little relation to general linear multistep methods, but will use them only as a convenient starting point. Later on we shall consider another possible way of deriving schemes of the form (2.51) and this removes any apparent relationship between these schemes and general linear multistep methods. This new method of derivation may be used to establish a close link, in certain cases, between schemes of the form (2.51) and the method of deferred correction developed by Fox (1957). In Chapter 3 we shall use particular integration schemes with t fixed at either 1 or 2 and this allows us to develop efficient procedures for the numerical integration of stiff systems. It will become clear from the analysis of the next chapter that it is possible to use schemes with t fixed at a larger value than two in order to obtain higher degrees of accuracy. This would seem to be another area which merits further research.

We could also use algorithms of this general class to generate solutions of recurrence relations which have not arisen as a result of applying a finite difference scheme to a system of O.D.E.s. In this case it is necessary to apply our iteration schemes in a slightly different way in that we iterate to convergence rather than performing a fixed number of primary iterations. As mentioned previously these iteration schemes must be used with care since they are not as widely applicable as iteration schemes used in previous sections. In cases where the required solution is slowly varying, however, we may often obtain rapid convergence using algorithms similar to (2.51). Rather than considering our iteration scheme any further we shall finish off this chapter by examining a particular example.

Example 2.6

Consider the numerical solution of the linear recurrence relation

$$y_{n+1} + b(n)y_n + c(n)y_{n-1} = 0, \qquad y_1 = 1, \tag{2.52}$$

where

$$b(n) = -\left\{\frac{n^4 + 22n^3 + 130n^2 + 111n}{n^3 + 10n^2 + 10n + 1}\right\},$$

$$c(n) = \frac{n^3 + 21n^2 + 121n + 110}{n^3 + 10n^2 + 10n + 1},$$

and where it is known from some external source that the required solution of (2.52) is the relatively slowly varying one. The general solution of (2.52) is given by

$$y_n = c_1/n + c_2(n + 10)!,$$

where the given conditions are clearly such that $c_1 = 1, c_2 = 0$. We consider the case where the solution of (2.52) is required correct to six decimal places in the range $2 \leq n \leq 10$ and we shall attempt to generate the required solution of (2.52) using the iteration scheme

$$[b(n) + 2]y_n^{(t)} + [c(n) - 1]y_{n-1}^{(t)} = -\delta^2 y_n^{(t-1)}, \qquad n = 2, 3, \ldots,$$

with $y_1^{(t)} \equiv 1$, for all $t \geq 0$, and $y_n^{(0)} \equiv 0$, for all $n \geq 2$.

We do not attempt to calculate our sequence of iterates in a sophisticated, space saving order but instead compute $\{y_n^{(1)}\}, \{y_n^{(2)}\}, \ldots$ in succession for all n in the range [2, 15] in the hope of obtaining convergence to the required degree of precision over the whole range [2, 10].

It can be seen from Table 2.6 that convergence to the prescribed degree of precision is obtained in about six iterations, with more accuracy generally being obtained as n increases. In fact one extra iteration would be required to obtain y_2 to the prescribed degree of precision. Otherwise the results obtained are satisfactory.

TABLE 2.6

n	1st Iteration	2nd Iteration	3rd Iteration	4th Iteration	5th Iteration	6th Iteration	True solution
2	0.4711055	0·5028043	0·4996773	0·5000402	0·4999948	0·5000007	0·5000000
3	0·3078957	0·3352190	0·3331514	0·3333535	0·3333309	0·3333336	0·3333333
4	0·2286913	0·2513358	0·2498812	0·2500127	0·2499985	0·2500002	0·2500000
5	0·1819267	0·2010109	0·1999134	0·2000092	0·1999989	0·2000001	0·2000000
6	0·1510603	0·1674711	0·1665990	0·1666739	0·1666658	0·1666668	0·1666666
7	0·1291612	0·1435214	0·1428016	0·1428631	0·1428564	0·1428572	0·1428571
8	0·1128159	0·1255640	0·1249529	0·1250051	0·1249994	0·1250001	0·1250000
9	0·1001484	0·1116003	0·1110702	0·1111156	0·1111106	0·1111112	0·1111111
10	0·0900423	0·1004316	0·0999638	0·1000039	0·9999995	0·1000001	0·1000000

3

Iterative Algorithms for the Solution of Differential Equations

In the previous chapter we outlined a theory pertaining to the stable generation of certain non-dominant solutions of linear recurrence relations. It was shown that in some cases by disposing with an appropriate number of the initial conditions it is possible to replace an ill-posed *initial value problem* by an *equivalent* well-posed *boundary value problem;* the equivalence being in the sense that the boundary value problem has a solution which, over some finite range, approximates to an arbitrary degree of precision the required solution of the original initial value problem. We were also able to show that in the non-linear case our original initial value problem could sometimes be re-posed as an equivalent boundary value problem and that this transformation could be effected via a precisely defined iterative scheme which in certain circumstances gave a rapidly convergent sequence of iterates which converged to the required solution without requiring the initial approximations to be particularly accurate. In the present chapter the above theory is applied to the numerical solution of initial value problems for ordinary differential equations using linear multistep methods. Discretisation algorithms are derived which in certain cases are not even zero-stable when solved directly in the usual way but which have large regions of absolute stability when the required solution is generated using one of the iterative techniques developed in the previous chapter. From this development we may draw the following important conclusion, which is really one of the key messages of this text. This is that it is not meaningful merely to state whether a linear k-step method used with a particular value of the steplength is stable or not. Instead the precise way in which the required solution is to be obtained must be stated and then, and only then, can the composite method (= linear multistep method + precise method of obtaining the required solution) be spoken of as being either stable or unstable. No problems of this sort have been encountered so far in developing the general theory of stability for discretisation schemes, since invariably it is tacitly assumed that the required solution is to be generated directly so that no such ambiguities can arise. If the algorithms to be developed in this chapter are to be used, however, fundamental changes in some of the basic stability definitions are required. The basis of such a theory of stability for iterative schemes will be given in Section 3.4.

Our main aim in developing this iterative approach is to derive efficient algorithms for the numerical solution of stiff systems of equations, since it is in this area that a totally new approach could be beneficial. However, for the sake of completeness, and since it is a convenient starting place, we begin our analysis with a brief discussion of the non-stiff case, although the existing theory for this particular case is relatively satisfactory and some excellent codes are now available.

3.1 Non-stiff systems of equations

In this section we are concerned primarily with the approximate numerical integration of the first order system

$$\frac{d\mathbf{y}}{dx} = \mathbf{f}(x, \mathbf{y}), \qquad \mathbf{y}(x_0) = \mathbf{y}_0 \tag{3.1}$$

using the linear k-step method

$$\sum_{j=0}^{k} \alpha_j \mathbf{y}_{n+j} = h \sum_{j=0}^{k} \beta_j \mathbf{f}_{n+j} + h^2 \sum_{j=0}^{k} \gamma_j \mathbf{g}_{n+j}, \tag{3.2}$$

where

$$\mathbf{f}_{n+j} = \mathbf{f}(x_{n+j}, \mathbf{y}_{n+j}) \quad \text{and} \quad \mathbf{g}_{n+j} = \frac{d}{dx}\mathbf{f}(x, \mathbf{y}(x))\Bigg|_{\substack{x = x_{n+j} \\ \mathbf{y} = \mathbf{y}_{n+j}}}$$

$$\equiv \mathbf{f}_x(x_{n+j}, \mathbf{y}_{n+j}) + \mathbf{f}_y(x_{n+j}, \mathbf{y}_{n+j})\mathbf{f}(x_{n+j}, \mathbf{y}_{n+j}).$$

In (3.1) it is assumed that the function $\mathbf{f}(x, \mathbf{y})$ satisfies the conditions necessary for a unique solution $\mathbf{y}(x)$ to exist. For the purposes of our future analysis we find it convenient to introduce the polynomials

$$\rho(v) = \sum_{i=0}^{k} \alpha_i v^i,$$

$$\sigma(v) = \sum_{i=0}^{k} \beta_i v^i,$$

$$\tau(v) = \sum_{i=0}^{k} \gamma_i v^i,$$

which are called respectively the *first*, *second* and *third characteristic polynomials* associated with (3.2). From these polynomials we may construct three new sequences of polynomials $\rho_j(v)$, $\sigma_j(v)$ and $\tau_j(v)$ defined by the relations

$$\rho_1(v) = \rho(v), \qquad \sigma_1(v) = \sigma(v), \qquad \tau_1(v) = \tau(v),$$

$$\rho_{j+1}(v) = v\rho_j'(v), \qquad \sigma_{j+1}(v) = v\sigma_j'(v),$$

$$\tau_{j+1}(v) = v\tau_j'(v), \qquad j \geq 1, \qquad \text{where } ' \equiv \frac{\mathrm{d}}{\mathrm{d}v}.$$

It may easily be verified that the linear multistep method (3.2) has exact order m if and only if

$$\left.\begin{array}{ll} \rho_1(1) = 0, \qquad \rho_2(1) = \sigma(1), & \\ \rho_{j+1}(1) = j\sigma_j(1) + j(j-1)\tau_{j-1}(1) & j = 2, 3, \ldots, m, \\ \text{and} & \\ \rho_{m+2}(1) \neq (m+1)\sigma_{m+1}(1) + m(m+1)\tau_m(1), & \end{array}\right\} \qquad (3.3)$$

and the principal term in the local truncation error associated with (3.2) is given by

$$[\rho_{m+2}(1) - (m+1)\sigma_{m+1}(1) - m(m+1)\tau_m(1)]\frac{h^{m+1}}{(m+1)!}\mathbf{y}^{(m+1)}. \qquad (3.4)$$

Relations (3.3) provide a convenient procedure for determining the order of any linear k-step scheme of the form (3.2) whilst the corresponding expression for the principal term in the local truncation error can be found from (3.4).

The numerical solution of (3.2) is to be effected using a precisely defined iterative technique rather than the more usual direct method. We therefore try to develop a class of high order integration procedures based on iterative schemes which converge rapidly to the required solution without requiring the initial approximations to be particularly accurate. In fact we now consider just one particular class of iteration schemes for the solution of (3.2). These take the general form

$$\alpha_k \mathbf{y}_{n+k}^{(t-1)} + \sum_{j=0}^{k-1} \alpha_j \mathbf{y}_{n+j}^{(t)} = h\beta_k \mathbf{f}(x_{n+k}, \mathbf{y}_{n+k}^{(t-1)})$$

$$+ h \sum_{j=0}^{k-1} \beta_j \mathbf{f}(x_{n+j}, \mathbf{y}_{n+j}^{(t)}) + h^2 \gamma_k \mathbf{g}(x_{n+k}, \mathbf{y}_{n+k}^{(t-1)})$$

$$+ h^2 \sum_{j=0}^{k-1} \gamma_j \mathbf{g}(x_{n+j}, \mathbf{y}_{n+j}^{(t)}). \tag{3.5}$$

As was mentioned in the previous chapter, there are numerous possible iteration schemes suitable for the generation of a solution of (3.2), some of which may be more efficient than (3.5). However, for the sake of brevity and also since algorithms based on the class of schemes defined by (3.5) have proved to be relatively efficient in practice, compared with algorithms based on the other classes of iteration schemes so far considered, attention will be confined to (3.5) from now on. If we were to think of the quantity h appearing in (3.5) as being everywhere "small" in some appropriate sense it would seem plausible that the most rapid convergence (if convergence occurs at all!) can be obtained by setting $\alpha_k = 0$, since then all terms with iteration index $t - 1$ are multiplied by h or h^2. Practical experience would also seem to indicate that this is indeed the case in general and so we first consider schemes of the form (3.5) with $\alpha_k = 0$. The scheme (3.5) is further normalised by setting $\alpha_{k-1} = 1$. In order that the nomenclature will be consistent with the previous chapter, iteration scheme (3.5), which has iteration parameter t, is referred to as the primary iteration scheme. If our original system (3.1) is non-linear, the case which we will regard as being of prime importance in practice, it will be necessary to use a secondary iteration scheme to allow us to solve (3.5) for $\mathbf{y}_{n+k-1}^{(t)}$. Since, by assumption, system (3.1) is non-stiff we may usually generate the required solution of equation (3.2) efficiently using a *direct iteration scheme* (the method of solution of the non-linear algebraic equation $y = \phi(y)$ using $y^{(t)} = \phi(y^{(t-1)})$, $t = 1, 2, 3, \ldots$, is referred to as direct iteration; cf. (2.25)) and so our composite algorithm reduces to a Gauss–Seidel – direct-iteration scheme of the type developed in Section 2.4. Practical experience in the solution of recurrence relations arising from applications other than in the solution of ordinary differential equations indicates that in general there is little advantage to be gained in iterating the secondary iteration scheme to convergence in each cycle of the primary scheme since the solution obtained will not in general be the required solution of the original recurrence relation. This is especially so for small values of t. In general a more efficient computational procedure is to perform just one secondary (direct) iteration starting with the initial iterate $\mathbf{y}_{n+k-1}^{(t-1)}$. The value obtained after this single secondary iteration is denoted by $\mathbf{y}_{n+k-1}^{(t)}$. If this procedure is adopted the complete iteration scheme in explicit form can be written as

$$\mathbf{y}_{n+k-1}^{(t)} = -\sum_{j=0}^{k-2} [\alpha_j \mathbf{y}_{n+j}^{(t)} - h\beta_j \mathbf{f}(x_{n+j}, \mathbf{y}_{n+j}^{(t)}) - h^2\gamma_j \mathbf{g}(x_{n+j}, \mathbf{y}_{n+j}^{(t)})]$$

$$+ h \sum_{j=k-1}^{k} [\beta_j \mathbf{f}(x_{n+j}, \mathbf{y}_{n+j}^{(t-1)}) + h\gamma_j \mathbf{g}(x_{n+j}, \mathbf{y}_{n+j}^{(t-1)})]. \tag{3.6}$$

As will be shown later, in the special case in which we are now interested, i.e. where our non-linear recurrence relation arises as a result of integrating an ordinary differential equation using a linear multistep method, we are usually able to predict the $\{\mathbf{y}_i^{(0)}\}$ so accurately that $\mathbf{y}_{n+k-1}^{(1)}$ is the final required solution. In order to use iteration scheme (3.6), and in fact to use any scheme of the general form (3.5), we first of all need $k-1$ *initial conditions* of the form

$$\mathbf{y}_n^{(t)} = \mathbf{y}_n, \qquad n = 0, 1, 2, \ldots, k-2; \qquad t \geq 0.$$

Thus, in a sense, scheme (3.6) may be regarded as being a $k-1$ step method and this prompts us to give the following definition of *stepnumber*.

Definition 3.1

An iterative numerical integration procedure based on a linear multistep method is said to have stepnumber k if, for the purposes of starting, it requires to be specified k initial conditions of the form $\mathbf{y}(x_i) = \mathbf{y}_i$, $i = 0(1)k-1$.

If our iteration scheme defines the usual direct method of solution of (3.2), i.e. if $\beta_k = \gamma_k = 0$, then Definition 3.1 reduces to the more familiar concept of stepnumber, see, for example, Lambert (1973). Scheme (3.6) in practice also requires that a sequence of *initial approximations* of the form

$$\mathbf{y}_n^{(0)} = \mathbf{s}(n) \qquad \text{for all } n \geq k-1$$

be available. Here $\mathbf{s}(n)$ is some sequence of known initial approximations usually built up by means of a predictor as the integration proceeds. It is important at this stage to distinguish clearly between the terms "initial conditions" and "initial approximations" since both are needed if (3.6) is to be used for practical computation. The initial conditions are the $k-1$ values $\mathbf{y}_n$, $n = 0(1)k-2$, needed to start any linear $(k-1)$-step method, whilst the initial approximations $\mathbf{y}_n^{(0)}$, $n \geq k-1$, are the values needed to start the iteration scheme (3.6). These two terms will be constrained to have these precise meanings for the remainder of this section.

Any theoretical analysis of iteration scheme (3.6) needs to include an examination of the convergence of the sequence of iterates generated by (3.6) to the solution of (3.1) for increasing t, and for various sequences $\mathbf{s}(n)$ of initial approximations as h decreases monotonically to zero. Sufficient conditions for the convergence of the solution generated by scheme (3.6) to the solution of (3.1) are given in the following theorem.

Theorem 3.1

Consider the linear multistep method (3.2) *where the coefficients are chosen so that* $\alpha_i \leq 0$ *for all* $i = 0(1)k-2$. *Suppose now that the two functions* $\mathbf{f} \equiv \mathbf{f}(x, \mathbf{y}(x))$ *and* $\mathbf{g} \equiv \mathbf{g}(x, \mathbf{y}(x))$ *satisfy Lipschitz conditions with constants* K_1 *and* K_2 *respectively so that, at any two points* $\mathbf{y}_1$ *and* $\mathbf{y}_2$ *in the solution space of* (3.1) *and for any point* x *in the range of integration, relations of the form*

$$\|\mathbf{f}(x, \mathbf{y}_1) - \mathbf{f}(x, \mathbf{y}_2)\| \leq K_1 \|\mathbf{y}_1 - \mathbf{y}_2\|$$

and

$$\|\mathbf{g}(x, \mathbf{y}_1) - \mathbf{g}(x, \mathbf{y}_2)\| \leq K_2 \|\mathbf{y}_1 - \mathbf{y}_2\|$$

hold. Further suppose that the sequence $\mathbf{s}(n)$ *of initial approximations is generated using a consistent, zero-stable predictor with a non-zero region of absolute stability and that the initial conditions* $\mathbf{y}_i$ *are such that* $\mathbf{y}_i = \mathbf{y}(x_i) + \boldsymbol{\delta}_i(h)$, $i = 0, 1, \ldots, k-2$, *where* $\|\boldsymbol{\delta}_i(h)\|/h \downarrow 0$ *as* $h \downarrow 0$ *for all* $i = 0(1)k-2$. *Finally we assume that scheme* (3.2) *is consistent and that the calculations are performed with a maximum round-off error* $h\boldsymbol{\varepsilon}(h)$ *at each step where* $\boldsymbol{\varepsilon}(h) \downarrow 0$ *as* $h \downarrow 0$ *and* $\|\boldsymbol{\varepsilon}(h)\|$ *increases monotonically as h increases. Then under these assumptions it follows that the solution generated by iteration scheme* (3.6) *converges to the solution of* (3.1) *over any finite interval* $\mathbf{I}$ *of length* $\mathbf{L}$ *as* $h \downarrow 0$ *in the sense that* $\|\mathbf{y}_n^{(t)} - \mathbf{y}(x_n)\| \downarrow 0$ *as* $h \downarrow 0$ *for all finite t and for all* $x_n = nh + x_0 \in \mathbf{I}$.

Proof

The true solution $\mathbf{y}(x_n)$ of (3.1) satisfies a relation of the form

$$\sum_{j=0}^{k-1} \alpha_j \mathbf{y}(x_{n+j}) = h \sum_{j=0}^{k} \beta_j \mathbf{f}(x_{n+j}, \mathbf{y}(x_{n+j}))$$

$$+ h^2 \sum_{j=0}^{k} \gamma_j \mathbf{g}(x_{n+j}, y(x_{n+j})) + \mathbf{T}_n, \tag{3.7}$$

where $\mathbf{T}_n$ is the local truncation error associated with (3.2) at the steppoint x_n. The numerical solution, $\mathbf{y}_n^{(t)}$, which is actually calculated as the tth approximation to $\mathbf{y}_n$, satisfies the relation

$$\mathbf{y}_{n+k-1}^{(t)} = -\sum_{j=0}^{k-2} [\alpha_j \mathbf{y}_{n+j}^{(t)} - h\beta_j \mathbf{f}(x_{n+j}, \mathbf{y}_{n+j}^{(t)}) - h^2\gamma_j \mathbf{g}(x_{n+j}, \mathbf{y}_{n+j}^{(t)})]$$
$$+ h \sum_{j=k-1}^{k} [\beta_j \mathbf{f}(x_{n+j}, \mathbf{y}_{n+j}^{(t-1)}) + h\gamma_j \mathbf{g}(x_{n+j}, \mathbf{y}_{n+j}^{(t-1)})]$$
$$+ \mathbf{R}_n^{(t)}, \tag{3.8}$$

where $\mathbf{R}_n^{(t)}$ is the round-off error arising during the calculation of $\mathbf{y}_{n+k-1}^{(t)}$. Defining an error vector $\mathbf{e}_n^{(t)}$ as

$$\mathbf{e}_n^{(t)} = \mathbf{y}_n^{(t)} - \mathbf{y}(\mathrm{x}_n),$$

subtracting (3.7) from (3.8) and using the fact that both **f** and **g** satisfy Lipschitz conditions we obtain

$$\|\mathbf{e}_{n+k-1}^{(t)}\| \leq \sum_{j=0}^{k-2} [-\alpha_j \|\mathbf{e}_{n+j}^{(t)}\| + h\|\beta_j \mathbf{e}_{n+j}^{(t)}\| K_1 + h^2 \|\gamma_j \mathbf{e}_{n+j}^{(t)}\| K_2]$$
$$+ h \sum_{j=k-1}^{k} [\|\beta_j \mathbf{e}_{n+j}^{(t-1)}\| K_1 + h\|\gamma_j \mathbf{e}_{n+j}^{(t-1)}\| K_2] + \phi, \tag{3.9}$$

where $\phi = \max_{n,t} \|\mathbf{T}_n - \mathbf{R}_n^{(t)}\|$ with $n \leq \mathbf{L}/h$ and where we assume that the number of iterations, t, is finite. In order to simplify our notation at this stage we set

$$r = \sum_{j=0}^{k-2} [hK_1|\beta_j| + h^2K_2|\gamma_j|],$$

$$s = \sum_{j=k-1}^{k} [K_1|\beta_j| + hK_2|\gamma_j|]$$

and define two new quantities Y and $\varepsilon_n^{(t)}$ as

$$Y = \max_n \|\mathbf{e}_n^{(0)}\|, \qquad \varepsilon_n^{(t)} = \max_{m \leq n} \|\mathbf{e}_{m+k-1}^{(t)}\|.$$

From relation (3.9) with $t = 1$ it follows that

$$\|\mathbf{e}_{n+k-1}^{(1)}\| \leq (1 + r)\varepsilon_{n-1}^{(1)} + hsY + \phi, \tag{3.9a}$$

where we note that

$$-\sum_{j=0}^{k-2} \alpha_j = 1$$

by virtue of consistency together with our assumption that $\alpha_i \le 0$, $i = 0(1)k - 2$. It will be proved by induction that

$$\|\mathbf{e}_{n+k-1}^{(1)}\| \le (1 + r)^{n+1}\delta + (hsY + \phi) \sum_{i=0}^{n} (1 + r)^i, \tag{3.10}$$

where $\delta = \max_{0 \le i \le k-2}\|\boldsymbol{\delta}_i(h)\|$. Since $\varepsilon_n^{(t)} = \max(\varepsilon_{n-1}^{(t)}, \|\mathbf{e}_{n+k-1}^{(t)}\|)$ it follows immediately that the sequence $\varepsilon_n^{(t)}$ is a non-decreasing function of n. Furthermore, since clearly $\varepsilon_{n-1}^{(1)} < (1 + r)\varepsilon_{n-1}^{(1)} + hsY + \phi$, equation (3.9a) gives that

$$\varepsilon_n^{(1)} \le (1 + r)\varepsilon_{n-1}^{(1)} + hsY + \phi.$$

Successive application of this relation gives

$$\varepsilon_n^{(1)} \le (1 + r)^n \varepsilon_0^{(1)} + \sum_{i=0}^{n-1} (1 + r)^i [hsY + \phi].$$

But $\varepsilon_0^{(1)} = \|\mathbf{e}_{k-1}^{(1)}\| \le (1 + r)\delta + hsY + \phi$ from (3.9) with $n = 0, t = 1$ and so we have

$$\varepsilon_n^{(1)} \le (1 + r)^{n+1}\delta + \sum_{i=0}^{n} (1 + r)^i [hsY + \phi]$$

and hence (3.10) follows immediately. It may also be shown by induction that if a quantity R is defined as

$$R = \sum_{i=0}^{u} (1 + r)^i,$$

where u is the number of intervals in the range of integration ($= \mathbf{L}/h$), which we have assumed to be finite, then

$$|\varepsilon_n^{(t)}| \le (hsR)^t Y + [\phi R + \delta(1 + r)^{u+1}] \sum_{i=0}^{t-1} (hsR)^i. \tag{3.11}$$

We now examine the behaviour of the r.h.s. of (3.11) as $h \to 0$. We have assumed that the sequence of initial approximations $\{\mathbf{y}_n^{(0)}\}$ is generated using an explicit convergent predictor and so it follows that $Y \to 0$ as $h \to 0$. If we now examine the asymptotic behaviour of ϕ as $h \to 0$, we note first of all that

$$\mathbf{T}_n - \mathbf{R}_n^{(t)} \approx \mathbf{C}_{m+1}h^{m+1} - h\boldsymbol{\varepsilon}(h) \quad \text{as} \quad h \to 0,$$

where $m \ge 1$ from the consistency of our method

$$\approx h\boldsymbol{\mu}(h) \quad \text{where} \quad \boldsymbol{\mu}(h) \to 0 \quad \text{as} \quad h \to 0.$$

It now follows immediately that $\phi/h \to 0$ as $h \to 0$. We also note that

$$hR = h \sum_{i=0}^{u} (1 + r)^i \to z(L) \quad \text{as} \quad h \to 0,$$

where $z(L)$ is bounded, and that $s \to K_1 [|\beta_{k-1}| + |\beta_k|]$ as $h \to 0$.

From this it follows that

$$(hsR)^t \quad \text{and} \quad hR \sum_{i=0}^{t-1} (hsR)^i$$

are both bounded for all finite t and that the number of iterations t is by assumption bounded. From (3.11) it now follows that $\|\boldsymbol{\varepsilon}_n^{(t)}\| \to 0$ as $h \to 0$ and hence $\|\mathbf{e}_n^{(t)}\| \to 0$ as $h \to 0$, so that the convergence of scheme (3.6) is proved.

It is important to note that the estimates which we have used in various parts of the proof of Theorem 3.1 have been rather crude and the bound on the r.h.s. of (3.11) is usually a gross overestimate of the error in our solution. In particular it should be noted that relation (3.11) does not guarantee that, for sufficiently small $h > 0$, the error $\mathbf{e}_n^{(t)}$ tends to zero, or even decreases, with increasing t. In practice, however, it does turn out that under favourable circumstances our error $\varepsilon_n^{(t)}$ does decrease with increasing t. This theoretical aspect of our algorithms will be examined more fully in Section 3.4.

3.2 Adams-type methods

One of the conditions under which we have established the convergence of our basic algorithm (3.6) is the requirement that all of the coefficients α_i, apart from α_{k-1}, should be less than or equal to zero. It may be shown (Copson, 1970) that this non-positivity assumption on the α_i is sufficient to ensure that our basic linear multistep method is zero-stable. One of the main classes of schemes in which we shall be interested are those with $\alpha_{k-1} = 1$, $\alpha_{k-2} = -1$, $\alpha_i = 0$, $i < k - 2$, i.e. *Adams type formulae*, since these have proved to be particularly efficient in practice especially for the purposes of starting our integration. If we consider first of all the case $\gamma_i \equiv 0$ in (3.2) we may list the following implicit Adams-type schemes:

$$\mathbf{y}_{n+1}^{(t)} - \mathbf{y}_n^{(t)} = h\{-\tfrac{1}{12}\mathbf{f}(x_{n+2}, \mathbf{y}_{n+2}^{(t-1)}) + \tfrac{2}{3}\mathbf{f}(x_{n+1}, \mathbf{y}_{n+1}^{(t)}) + \tfrac{5}{12}\mathbf{f}(x_n, \mathbf{y}_n^{(t)})\} \tag{3.12}$$

Order = 3, error constant = 1/24, stepnumber = 1.

$$\mathbf{y}_{n+2}^{(t)} - \mathbf{y}_{n+1}^{(t)} = \frac{h}{24}\{-\mathbf{f}(x_{n+3}, \mathbf{y}_{n+3}^{(t-1)}) + 13\mathbf{f}(x_{n+2}, \mathbf{y}_{n+2}^{(t)})$$
$$+ 13\mathbf{f}(x_{n+1}, \mathbf{y}_{n+1}^{(t)}) - \mathbf{f}(x_n, \mathbf{y}_n^{(t)})\} \quad (3.13)$$

Order = 4, error constant = 11/720, stepnumber = 2.

If we now consider the case $\gamma_i \neq 0$ we may derive the following one-step scheme:

$$\mathbf{y}_{n+1}^{(t)} - \mathbf{y}_n^{(t)} = h\{\tfrac{1}{120}\mathbf{f}(x_{n+2}, \mathbf{y}_{n+2}^{(t-1)}) + \tfrac{8}{15}\mathbf{f}(x_{n+1}, \mathbf{y}_{n+1}^{(t)}) + \tfrac{11}{24}\mathbf{f}(x_n, \mathbf{y}_n^{(t)})\}$$
$$+ h^2\{-\tfrac{7}{60}\mathbf{g}(x_{n+1}, \mathbf{y}_{n+1}^{(t)}) + \tfrac{1}{15}\mathbf{g}(x_n, \mathbf{y}_n^{(t)})\} \quad (3.14)$$

Order = 5, error constant = −1/2400, stepnumber = 1.

It is quite clear that it is possible to derive even higher order schemes belonging to these two classes defined by the cases $\gamma_i \equiv 0$ and $\gamma_i \neq 0$ by considering increased values of the integer k. However, we shall not do so in this text since, once the way in which schemes (3.12)–(3.14) are to be used in practice is well understood, the method of derivation and use of higher order schemes will follow immediately. Scheme (3.14) will be mainly of use for problems which are such that the required second derivatives are reasonably cheap to evaluate, so that there is some gain in computational efficiency to be had by using formulae employing second derivatives. Practical experience has shown that probably the most efficient way of using Adams methods of the type (3.12)–(3.14) is to use them as correctors with lower order conventional Adams methods being used as predictors. For the time being we shall confine our attention to Scheme (3.12) since the way in which the other two schemes are used is very similar.

Suppose now that we have a particular problem of the form (3.1) to integrate and that we are going to attempt to generate the required solution using a (variable order–variable step) Adams method. We shall in fact consider three variations of conventional Adams schemes, each of which has a certain desirable property. In the first scheme an explicit first order Adams method is used as the predictor whilst for the corrector we use an implicit second order Adams method. These are used in PECE (*P*redict, *E*valuate, *C*orrect, *E*valuate) mode in the way suggested by Shampine and Gordon (1975). It is assumed that a steplength h which is suitable for starting our problem has been found and that it is required to calculate a numerical approximation to $\mathbf{y}(x_i)$ at the grid points $x_i = x_0 + ih$. The following PECE scheme may be used to generate second order approximations at both x_1 and x_2:

$$
\begin{aligned}
\mathbf{f}_0 &= \mathbf{f}(x_0, \mathbf{y}_0),\\
\mathbf{y}_1^p &= \mathbf{y}_0 + h\mathbf{f}_0,\\
\mathbf{f}_1^p &= \mathbf{f}(x_1, \mathbf{y}_1^p),\\
\mathbf{y}_1^c &= \mathbf{y}_0 + \frac{h}{2}(\mathbf{f}_0 + \mathbf{f}_1^p),\\
\mathbf{f}_1^c &= \mathbf{f}(x_1, \mathbf{y}_1^c),\\
\mathbf{y}_2^p &= \mathbf{y}_1^c + h\mathbf{f}_1^c,\\
\mathbf{f}_2^p &= \mathbf{f}(x_2, \mathbf{y}_2^p),\\
\mathbf{y}_2^c &= \mathbf{y}_1^c + \frac{h}{2}(\mathbf{f}_1^c + \mathbf{f}_2^p),\\
\mathbf{f}_2^c &= \mathbf{f}(x_2, \mathbf{y}_2^c).
\end{aligned}
\tag{3.14a}
$$

If we continue to use our Adams scheme in PECE mode we see that two function evaluations are required at all acceptable steps (i.e. where estimated local truncation error is less than a prescribed maximum), apart from the first. As mentioned previously we shall be concerned only with problems for which function evaluations are very expensive compared with the other computations involved. For problems of this type the number of function evaluations is a good measure of the total amount of work required, since problems for which function evaluations are cheap can generally be integrated by almost any standard method using a small amount of computation. The number of function evaluations required per step may be halved if we do not compute $\mathbf{f}_1^c$ and all the subsequent calls of $\mathbf{f}_1^c$ in our code are replaced by $\mathbf{f}_1^p$ and similarly for $\mathbf{f}_2^c$, etc. It turns out, however, that there is a loss in accuracy and a significant decrease in the stability region if this new approach is adopted, so that generally it is not worthwhile to make these modifications. If, however, we return to our scheme used in PECE mode we find that we are able to obtain a third order approximation at the point x_1 by using scheme (3.12). Thus we obtain

$$\mathbf{y}_1^{cc} = \mathbf{y}_0 + h\{-\tfrac{1}{12}\mathbf{f}_2^c + \tfrac{2}{3}\mathbf{f}_1^c + \tfrac{5}{12}\mathbf{f}_0\},$$

where $\mathbf{y}_1^{cc}$ is taken as the final third order approximation at the point x_1. Virtually all the practical facilities suggested by Shampine and Gordon (1975) can be used with this new integration procedure and, in particular, the difference $\mathbf{y}_1^{cc} - \mathbf{y}_1^c$ serves as a computable estimate of the local truncation error in $\mathbf{y}_1^c$. If this error is less than some prescribed tolerance, local extrapolation is performed and $\mathbf{y}_1^{cc}$ is taken as the finally accepted approximation at x_1. If the stepsize remains unchanged, we may continue to use our procedure in the following way:

$$\mathbf{y}_3^p = \mathbf{y}_2^c + h\mathbf{f}_2^c,$$

$$\mathbf{f}_3^p = \mathbf{f}(x_3, \mathbf{y}_3^p),$$

$$\mathbf{y}_3^c = \mathbf{y}_2^c + \frac{h}{2}\{\mathbf{f}_2^c + \mathbf{f}_3^p\},$$

$$\mathbf{f}_3^c = \mathbf{f}(x_3, \mathbf{y}_3^c)$$

$$\mathbf{y}_2^{cc} = \mathbf{y}_1^{cc} + h\{-\tfrac{1}{12}\mathbf{f}_3^c + \tfrac{2}{3}\mathbf{f}_2^c + \tfrac{5}{12}\mathbf{f}_1^c\},$$

where we note in particular that there is no need to evaluate $\mathbf{f}(x_1, \mathbf{y}_1^{cc})$. If this procedure is now continued with a fixed stepsize, it can be seen that it requires two function evaluations per step but that it always has degree one higher than the standard one-step Adams method used in PECE mode. Since at each step we are effectively using one predictor and two different correctors we shall refer to the scheme just described as a one-step Adams method used in PEC_1EC_2 mode.

If we apply this scheme to our usual scalar test equation $y' = \lambda y$, where λ is a complex constant with negative real part, we obtain a rather complicated expression, which changes at every step, for the characteristic root. It may be shown, however, that the region of absolute stability of our scheme is very similar to that of the first order Adams predictor–corrector scheme used in PECE mode (Shampine and Gordon, 1975), and in particular that the intercepts on the negative real axis are $(0, -2)$. If the steplength of integration is changed at any point to $\bar{h} = \omega h$ it will be necessary to modify the second corrector for the computation of $\mathbf{y}_2^{cc}$ to

$$\mathbf{y}_n - \mathbf{y}_{n-1} = h\{r\mathbf{y}'_{n+\omega} + s\mathbf{y}'_n + t\mathbf{y}'_{n-1}\}, \tag{3.15}$$

where

$$r = -\frac{1}{6(\omega + \omega^2)}, \; s = \frac{1 + 4\omega + 3\omega^2}{6(\omega + \omega^2)}, \; t = \frac{2\omega + 3\omega^2}{6(\omega + \omega^2)}.$$

The way in which the stepsize is changed is slightly different from usual and may best be described as follows. Suppose that the finally accepted approximation to the solution at the point x_{i-1} has been computed and we wish to calculate an approximate solution at x_i. This may be done using the PEC_1EC_2 scheme just described and an estimate $\mathbf{y}_i^{cc} - \mathbf{y}_i^c$ of the local truncation error in $\mathbf{y}_i^c$ is readily available. Suppose now that on the basis of this error estimate it is decided to change the steplength to ωh. Then if $\omega < 1$ it will be necessary to go back to the steppoint x_{i-1}, in order to calculate second order approximations at $x_{i-1} + \omega h, x_{i-1} + 2\omega h$ using our PEC_1E scheme and finally to calculate a third order approximation at $x_{i-1} + \omega h$ using our second corrector. This is of course very similar to what is usually done in practice when using single-step schemes. If $\omega > 1$, however, we change the step $[x_{i+1}, x_{i+2}]$ to ωh (and keep the step $[x_i, x_{i+1}]$ fixed!) and then use scheme (3.15) rather than (3.12) as the second corrector. If this approach is adopted the next solution is computed at $x_{i+1} = x_i + h$ and so in effect the stepchange is applied after a 'delay' of one step. It is clear that this procedure will allow a saving of two function evaluations every time that a step increase is performed and practical experience has shown that this delay in changing the step does not cause any additional difficulties. Since there are effectively two different correctors and one predictor there are several variants of this approach. We will denote the scheme just described by Scheme 1. Another possible scheme, which we denote by Scheme 2, and which is also of third order and requires only two function evaluations per step, is as follows:

$$\begin{aligned}
\mathbf{f}_0^c &= \mathbf{f}(x_0, y_0).\\
\mathbf{y}_1^p &= \mathbf{y}_0 + h\mathbf{f}_0^c,\\
\mathbf{f}_1^p &= \mathbf{f}(x_1, \mathbf{y}_1^p),\\
\mathbf{y}_1^c &= \mathbf{y}_0 + \frac{h}{2}(\mathbf{f}_0^c + \mathbf{f}_1^p),\\
\mathbf{f}_1^c &= \mathbf{f}(x_1, \mathbf{y}_1^c),\\
\mathbf{y}_2^p &= \mathbf{y}_1^c + h\{\tfrac{3}{2}\mathbf{f}_1^c - \tfrac{1}{2}\mathbf{f}_0^c\},\\
\mathbf{f}_2^p &= \mathbf{f}(x_2, \mathbf{y}_2^p),\\
\mathbf{y}_1^{cc} &= \mathbf{y}_0 + h\{-\tfrac{1}{12}\mathbf{f}_2^p + \tfrac{2}{3}\mathbf{f}_1^c + \tfrac{5}{12}\mathbf{f}_0^c\},\\
\mathbf{y}_2^c &= \mathbf{y}_1^{cc} + h\{\tfrac{5}{12}\mathbf{f}_2^p + \tfrac{2}{3}\mathbf{f}_1^c - \tfrac{1}{12}\mathbf{f}_0^c\},\\
\mathbf{f}_2^c &= \mathbf{f}(x_2, \mathbf{y}_2^c).
\end{aligned} \tag{3.15a}$$

An approximation to the error in $\mathbf{y}_1^c$ is given by $\mathbf{y}_1^{cc} - \mathbf{y}_1^c$ and an approximation to the error in $\mathbf{y}_2^p$ is $\mathbf{y}_2^c - \mathbf{y}_2^p$. Generally no action is taken on the error estimate obtained at x_1 (although if a catastrophic error is indicated here by a large difference between $\mathbf{y}_1^p$ and $\mathbf{y}_1^c$ we do not complete the cycle but instead start again from x_0 with a reduced stepsize). All decisions regarding the acceptability of the step from x_0 to x_2, and the parameters to be chosen for the next integration step, are made at x_2. Thus we may regard our scheme as being a block integration method whereby the integration is advanced two steps at a time. We have found it convenient in practice to use local extrapolation as suggested by Shampine and Gordon so that the estimates relate to those errors committed in $\mathbf{y}_1^c$ and $\mathbf{y}_2^p$ whereas the values $\mathbf{y}_1^{cc}$ and $\mathbf{y}_2^c$ are those which are actually carried forward. Thus, provided the error estimate at x_2 is less than a prescribed tolerance, $\mathbf{y}_1^{cc}$, $\mathbf{y}_2^c$ are taken as the finally accepted approximations at x_1 and x_2 respectively. If we apply Scheme 2 to the scalar test equation $y' = \lambda y$ and set $q = h\lambda$ we obtain

$$y_1^{cc} = \left(1 + q + \frac{q^2}{2} + \frac{q^3}{6} - \frac{q^4}{16}\right) y_0$$

and

$$y_2^c = (1 + 2q + 2q^2 + \tfrac{8}{6}q^3 + \tfrac{1}{4}q^4) y_0.$$

Since y_1^{cc} may be regarded as being an intermediate value in the middle of our block, we shall regard our scheme as being stable for those values of q which are such that

$$\left|\frac{y_2^c}{y_0}\right| \leq 1.$$

The problem is now to find this region of stability.

This may easily be done using a numerical method and the corresponding region of absolute stability is shown in Fig. 3.1. It can be seen that the stability region is somewhat smaller than that of the classic second order Adams predictor–corrector pair although it is still of a reasonable size.

Scheme 2 may now be continued in the following straightforward fashion:

$$\mathbf{y}_3^p = \mathbf{y}_2^c + h\mathbf{f}_2^c,$$

$$\mathbf{f}_3^p = \mathbf{f}(x_3, \mathbf{y}_3^p),$$

$$\mathbf{y}_3^c = \mathbf{y}_2^c + \frac{h}{2}(\mathbf{f}_2^c + \mathbf{f}_3^p),$$

$$\mathbf{f}_3^c = \mathbf{f}(x_3, \mathbf{y}_3^c),$$

$$\mathbf{y}_4^p = \mathbf{y}_3^c + h\{\tfrac{3}{2}\mathbf{f}_3^c - \tfrac{1}{2}\mathbf{f}_2^c\},$$

$$\mathbf{f}_4^p = \mathbf{f}(x_4, \mathbf{y}_4^p),$$

$$\mathbf{y}_3^{cc} = \mathbf{y}_2^c + h\{-\tfrac{1}{12}\mathbf{f}_4^p + \tfrac{2}{3}\mathbf{f}_3^c + \tfrac{5}{12}\mathbf{f}_2^c\},$$

$$\mathbf{y}_4^c = \mathbf{y}_3^{cc} + h\{\tfrac{5}{12}\mathbf{f}_4^p + \tfrac{2}{3}\mathbf{f}_3^c - \tfrac{1}{12}\mathbf{f}_2^c\},$$

$$\mathbf{f}_4^c = \mathbf{f}(x_4, \mathbf{y}_4^c),$$

and similarly over the whole range of integration. We note in particular that changes in the steplength of integration are very easy to perform when using this scheme and, since it is one-step in nature, it has proved to be very useful for starting purposes.

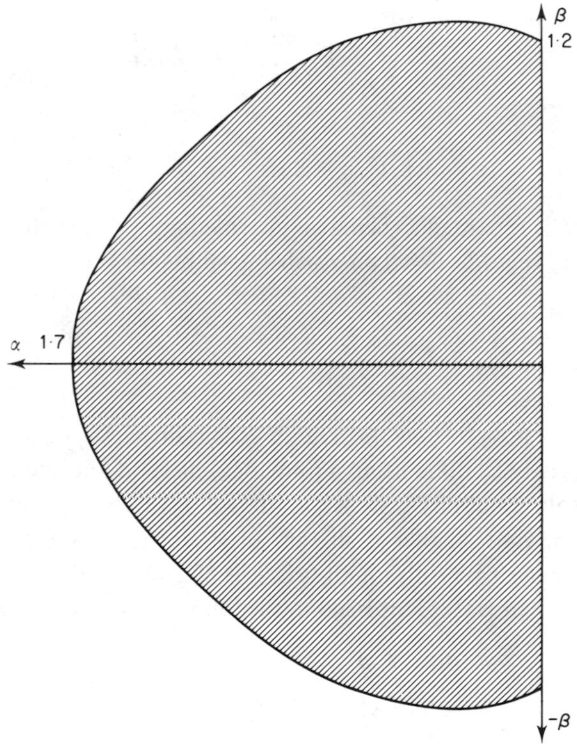

FIG. 3.1 Stability region (shaded) of Scheme 2.

It is also possible to derive integration procedures which incorporate scheme (3.12), which require extra function evaluations per step but which are designed to have increased regions of absolute stability. As an example we consider the following scheme which we denote by Scheme 3:

$$
\begin{aligned}
\mathbf{f}_0^c &= \mathbf{f}(x_0, \mathbf{y}_0),\\
\mathbf{y}_1^p &= \mathbf{y}_0 + h\mathbf{f}_0^c,\\
\mathbf{f}_1^p &= \mathbf{f}(x_1, \mathbf{y}_1^p),\\
\mathbf{y}_2^p &= \mathbf{y}_1^p + h\mathbf{f}_1^p,\\
\mathbf{f}_2^p &= \mathbf{f}(x_2, \mathbf{y}_2^p),\\
\mathbf{y}_1^c &= \mathbf{y}_0 + h\{r\mathbf{f}_2^p + (\tfrac{1}{2} - 2r)\mathbf{f}_1^p + (r + \tfrac{1}{2})\mathbf{f}_0^c\},
\end{aligned}
\tag{3.15b}
$$

where r is at this stage an undetermined parameter. It may be shown that this scheme has order 2 for any choice of r. Applying our scheme to the scalar test equation $y' = \lambda y$ we obtain

$$y_1^c = \left(1 + q + \frac{q^2}{2} + rq^3\right) y_0. \tag{3.16}$$

We may now choose the value of r so as to give our scheme as large a region of absolute stability as possible. By using a numerical method it was found that the required value of r is $r = 1/16$ which produces the scheme

$$y_1^c = y_0 + h\{\tfrac{1}{16}\mathbf{f}_2^p + \tfrac{3}{8}\mathbf{f}_1^p + \tfrac{9}{16}\mathbf{f}_0^c\},$$

which has the region of absolute stability shown in Fig. 3.2. We note in

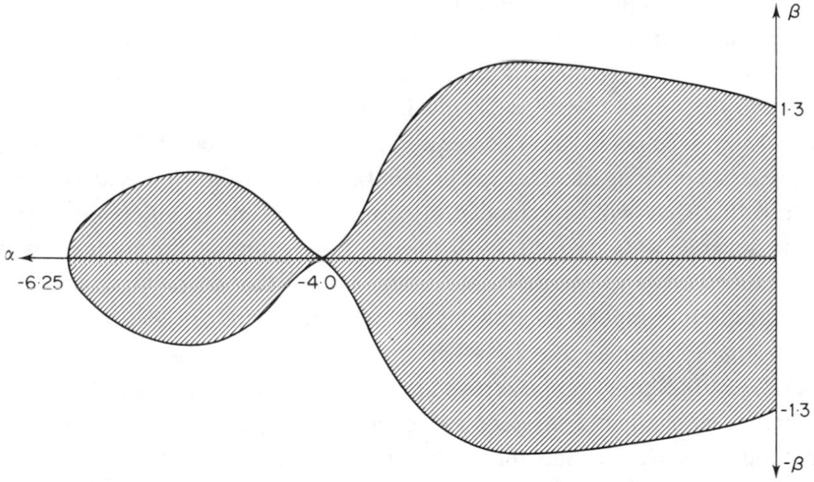

FIG. 3.2 Stability region (shaded) of Scheme 3.

particular that the stability region cuts the negative real axis at $(0, -6.25)$ and this is more than twice the interval of absolute stability possessed by any of the conventional Adams schemes used in PECE mode. The way in which this scheme may be continued is now clear. If the stepsize remains unchanged we have

$$\mathbf{f}_1^c = \mathbf{f}(x_1, \mathbf{y}_1^c),$$

$$\mathbf{y}_2^p = \mathbf{y}_1^c + h\mathbf{f}_1^c,$$

$$\mathbf{f}_2^p = \mathbf{f}(x_2, \mathbf{y}_2^p),$$

$$\mathbf{y}_3^p = \mathbf{y}_2^p + h\mathbf{f}_2^p,$$

$$\mathbf{f}_3^p = \mathbf{f}(x_3, \mathbf{y}_3^p),$$

$$\mathbf{y}_2^c = \mathbf{y}_1^c + h\{\tfrac{1}{16}\mathbf{f}_3^p + \tfrac{3}{8}\mathbf{f}_2^p + \tfrac{9}{16}\mathbf{f}_1^c\},$$

It is also clear that this scheme can be easily modified to take account of the case where the stepsize is changed at any point. At each steppoint the difference $\mathbf{y}_i^c - \mathbf{y}_i^p$ serves as a computable estimate of the local truncation error committed in $\mathbf{y}_i^p$. If this estimate is less than some pre-determined local tolerance, we shall as usual perform local extrapolation and carry the estimate $\mathbf{y}_i^c$ forward. Although this scheme requires one more function evaluation per step than the classic first order Adams scheme used in PECE mode, it has a stability interval more than three times as large. It should therefore be relatively efficient for problems, such as mildly stiff ones, where a large region of absolute stability is an important advantage.

It is important to note that all three of these approaches, (3.14a), (3.15a) and (3.15b), may be implemented using higher order schemes as well. For the sake of clarity we have considered only fixed order schemes which behave like single-step schemes in that they require only one initial condition. We are not, however, proposing them for use as general purpose algorithms although it is hoped that they will be extended at some future time to allow much higher order schemes to be derived. With Scheme 1, for example, the predictor will be an explicit Adams method of order $k - 1$, the corrector will be the corresponding implicit Adams method of order k and the scheme providing the value $\mathbf{y}_i^{cc}$ will be the $(k + 1)^{\text{th}}$ order method of the type defined by (3.12)–(3.13). Similar remarks also apply to Schemes 2 and 3, but for the sake of brevity we shall not consider any higher order schemes in this text since their method of derivation and their use is now clear.

Finally, while considering Adams methods of this class, we show how higher order schemes may be derived via the theory of interpolation. If we consider again the scalar equation

$$y' = f(x, y),$$

we have on integration that

$$y(x_{n+1}) - y(x_n) = \int_{x_n}^{x_{n+1}} f(x, y(x))\,dx. \tag{3.16a}$$

We now let $P(x)$ be the unique polynomial of degree 2 passing through the three points (x_n, f_n), (x_{n+1}, f_{n+1}) and (x_{n+2}, f_{n+2}). Then using the Newton form of the interpolating polynomial we have

$$P(x) = P(x_n + rh) = f_n + r(f_{n+1} - f_n) + \frac{r(r-1)}{2}(f_{n+2} - 2f_{n+1} + f_n).$$

We now make the approximation

$$\int_{x_n}^{x_{n+1}} f(x, y)\,dx \approx \int_{x_n}^{x_{n+1}} P(x)\,dx$$

so that in (3.16a)

$$\begin{aligned} y_{n+1} - y_n &= h\int_0^1 \left[f_n + r(f_{n+1} - f_n) + \frac{r(r-1)}{2}(f_{n+2} - 2f_{n+1} + f_n) \right] dr, \\ &= h[f_n + \tfrac{1}{2}(f_{n+1} - f_n) - \tfrac{1}{12}(f_{n+2} - 2f_{n+1} + f_n)], \\ &= h\{-\tfrac{1}{12}f_{n+2} + \tfrac{2}{3}f_{n+1} + \tfrac{5}{12}f_n\}, \end{aligned}$$

which is precisely scheme (3.12). It can also be verified that higher order schemes may be obtained in this way by interpolating the function f at the forward point x_{n+2}, the current point x_{n+1} and the backward points x_n, $x_{n-1}, \ldots, x_{n-p}$ and that the order of the scheme derived is $p + 3$.

We now present some numerical results to compare the performance of Schemes 1, 2 and 3 described earlier with that of the classic first order Adams pair used in PECE mode.

The first numerical experiment which we performed was to run all four methods for twenty integration steps on the linear problem.

$$\begin{aligned} y_1' &= y_2 \qquad & y_1(0) = 0 \\ y_2' &= -y_1 \qquad & y_2(0) = 1 \end{aligned}$$

and the maximum global truncation error committed over the given range for (i) $h = 0{\cdot}1$ and (ii) $h = 0{\cdot}01$ was found. What this experiment does is to give us some idea of the relative accuracies of the four methods for a given fixed stepsize. In Table 3.1 we present the results obtained for the solution of this problem and, as can be seen, Schemes 1 and 2 are more accurate than Scheme 3 and the classic Adams scheme. This confirms our expectations since Schemes 1 and 2 are of order 3 whereas the other two are only of order 2. The results actually given in Table 3.1 are for the L_∞-norm of the global truncation error and this was found to increase gradually as the integration proceeded.

TABLE 3.1

Adams	
$h = 0{\cdot}01$	Max. error $= 0{\cdot}327 \times 10^{-5}$
$h = 0{\cdot}1$	Max. error $= 0{\cdot}312 \times 10^{-2}$
Scheme 1	
$h = 0{\cdot}01$	Max. error $= 0{\cdot}18 \times 10^{-6}$
$h = 0{\cdot}1$	Max. error $= 0{\cdot}18 \times 10^{-2}$
Scheme 2	
$h = 0{\cdot}01$	Max. error $= 0{\cdot}40 \times 10^{-7}$
$h = 0{\cdot}1$	Max. error $= 0{\cdot}39 \times 10^{-3}$
Scheme 3	
$h = 0{\cdot}01$	Max. error $= 0{\cdot}20 \times 10^{-5}$
$h = 0{\cdot}1$	Max. error $= 0{\cdot}19 \times 10^{-2}$

As our second experiment we consider the numerical integration of the non-linear system

$$y_1' = -2y_1 + y_2^2 \qquad y_1(0) = 1$$

$$y_2' = y_1 y_2 - 1 \qquad y_2(0) = 0$$

integrated with a constant step $h = 0{\cdot}01$. In Table 3.2 we present the results obtained for the solution of this problem at $x = 0{\cdot}05$, $0{\cdot}1$.

As can be seen from Table 3.2 the same conclusion can be drawn for this non-linear problem as for the previous problem in that Schemes 1 and 2 are the most accurate of the four schemes considered.

Finally, as a third experiment we consider the numerical integration of a mildly stiff linear problem to illustrate the stability properties of our schemes. The problem chosen was

$$\begin{aligned} y_1' &= y_2 & y_1(1) &= \exp(-1) \\ y_2' &= -101y_2 - 100y_1 & y_2(1) &= -y_1(1) \end{aligned} \qquad x \geq 1.$$

The eigenvalues of the Jacobian matrix of this system are -1, -100 and so the system is mildly stiff. In Table 3.3 we present the results obtained for the component y_2 when this system is integrated using a constant step $h = 0{\cdot}04$. A

TABLE 3.2

	True solution	Adams	Scheme 1	Scheme 2	Scheme 3
$x = 0{\cdot}05$	$y_1 = 0{\cdot}9048795486$	$0{\cdot}9048865342$	$0{\cdot}9048792758$	$0{\cdot}9048794043$	$0{\cdot}9048848409$
	$y_2 = -0{\cdot}5118834596 \times 10^{-1}$	$-0{\cdot}5118584945 \times 10^{-1}$	$-0{\cdot}5118830681 \times 10^{-1}$	$-0{\cdot}5118833787 \times 10^{-1}$	$-0{\cdot}5118492250 \times 10^{-1}$
$x = 0{\cdot}10$	$y_1 = 0{\cdot}8190706166$	$0{\cdot}8190834945$	$0{\cdot}8190697070$	$0{\cdot}8190703307$	$0{\cdot}8190804929$
	$y_2 = -0{\cdot}1045146223$	$-0{\cdot}1045096572$	$-0{\cdot}1045144958$	$-0{\cdot}1045146104$	$-0{\cdot}1045078346$

similar behaviour was also exhibited by the component y_1 so y_1 is not listed in Table 3.3. The integration was started at $x = 1{\cdot}0$ so as to be outside the transient phase, where a small value of h is necessary to preserve accuracy.

It can be seen from the results of Table 3.3 that both the Adams method and Scheme 1 are unstable for this problem. By integrating further it was found that instability also occurs with Scheme 2, whereas Scheme 3 remains stable. All these results confirm our theoretical analysis.

It is worth emphasising again at this point that we are not proposing Schemes 1, 2 and 3 as general purpose algorithms. Rather we are introducing them as examples of a new approach which is based on the analysis of Chapter 2 and which seems worth investigating. The purpose of our numerical experiments is to show that our theoretical conclusions are indeed translated into practice, in that Schemes 1 and 2 are the more accurate of the schemes considered and Scheme 3 has superior stability properties. In practice the superior accuracy of Schemes 1 and 2 would result in them taking a larger stepsize than conventional Adams methods to achieve any prescribed degree of precision. As a result, for strict error tolerances at least, Schemes 1 and 2 generally require fewer function evaluations than the first order Adams pair over any given range of integration.

3.3 Some other classes of integration formulae

We now mention the possibility of deriving schemes which are such that $\alpha_{k-1} = 1$, $\alpha_i \leq 0$, $i < k-1$, but which are not of Adams type for the integration of (3.1). For any given stepnumber $k > 1$ these schemes have higher orders of accuracy than conventional Adams methods. One such scheme which is of order 5 is given by

$$\mathbf{y}_{n+2} - \tfrac{24}{57}\mathbf{y}_{n+1} - \tfrac{33}{57}\mathbf{y}_n = h\{-\tfrac{1}{57}\mathbf{f}_{n+3} + \tfrac{24}{57}\mathbf{f}_{n+2} + \mathbf{f}_{n+1} + \tfrac{10}{57}\mathbf{f}_n\}.$$

We shall not investigate this class of schemes any further in this text, beyond merely pointing out their existence, since once sufficient solution values are known their method of use is very similar to that just described for Adams methods.

Also in this section we note that there is the possibility of developing schemes which are not of Adams type and which are not used in PECE mode, but which are instead *iterated to convergence*. These schemes are usually not, in

TABLE 3.3

x	True solution	Adams method	Scheme 1	Scheme 2	Scheme 3
1·4	−0·246596939	−0·2466240813	−0·246593007	−0·2465955752	−0·246138036
1·8	−0·1652988882	−0·2677936127	$-0{\cdot}7488369766 \times 10^{-1}$	−0·1652970280	−0·1653214649
2·2	−0·1108158508	*	*	−0·1108023116	−0·1108258595

* Denotes overflow.

their present form, as efficient for non-stiff equations as the Adams-type schemes just described. However, as will be seen in Section 3.5 they form the basis for methods which are efficient for the numerical solution of stiff systems of equations so we find it convenient to introduce them here. Two explicit schemes which typify this approach are

$$\mathbf{y}_{n+2}^{(t-1)} - 4\mathbf{y}_{n+1}^{(t)} + 3\mathbf{y}_n^{(t)} = -2h\mathbf{f}(x_n, \mathbf{y}_n^{(t)}) \tag{3.17}$$

Order = 2, error constant = 2/3, stepnumber = 1.

$$\begin{aligned} y_{n+3}^{(t-1)} - 6y_{n+2}^{(t)} + 3\mathbf{y}_{n+1}^{(t)} + 2\mathbf{y}_n^{(t)} & \\ = -6h\mathbf{f}(x_{n+1}, \mathbf{y}_{n+1}^{(t)}) & \end{aligned} \tag{3.18}$$

Order = 3, error constant = 1/2, stepnumber = 2.

It can be seen from the examples (3.12)–(3.14) that we have given (and it would seem to be true in general) that, as with conventional linear multistep methods, the implicit Adams schemes have higher orders of accuracy and smaller error constants than the corresponding explicit schemes of type (3.17), (3.18) for a given stepnumber k. The method of use of schemes (3.17), (3.18) follows immediately from the analysis of Chapter 2. If an explicit predictor, such as Euler's method, is used to generate the initial sequence $\{\mathbf{y}_n^{(0)}\}$, we may generate a solution of (3.17) or (3.18) by the use of one of the iteration schemes (2.16) or (2.37b) iterated to convergence. Assuming that the iteration scheme does converge we are now immediately faced with the problem of whether the solution which we obtain on convergence is a good approximation to the solution of (3.1). Since our schemes are naturally all chosen to be consistent, it is the stability properties of the schemes which need investigation. This aspect will be examined in Section 3.4. It is worth pointing out at this stage that the implicit schemes (3.12)–(3.14) need not be used in PEC_1EC_2 mode, but may instead be corrected to convergence. We note in particular that scheme (3.13) is symmetric about the point $x_{n+(3/2)}$ and a Taylor series expansion about this point reveals that (3.13) has a local truncation error with an asymptotic expansion containing only even powers of h. In order to be able to use this scheme for practical computation we need two initial conditions of the form

$$\mathbf{y}(x_n) = \mathbf{y}_n, \qquad n = 0, 1,$$

to be specified initially. It is plausible that if the extra "initial" condition at the point x_1 were to be chosen in an appropriate manner (Stetter, 1973, p. 232) the resulting scheme would have a global truncation error with an asymptotic expansion containing only even powers of h. This would then form a firm basis for a procedure based on Richardson extrapolation. For systems which are moderately stiff, and so need to be integrated using procedures possessing more than just a very small region of absolute stability, a procedure which is based on (3.13) and which uses a suitable secondary iteration scheme may well be useful.

We now mention briefly a somewhat similar approach to that discussed earlier in this section. This has been developed for the non-stiff case by Urabe (1970), but has not so far been extended to deal with stiff systems. Urabe proposed using the integration procedure

$$\mathbf{y}_{n+1} - \mathbf{y}_n = \frac{h}{240}\{101\mathbf{f}_n + 128\mathbf{f}_{n+1} + 11\hat{\mathbf{f}}_{n+2}\} + \frac{h^2}{240}\{13\mathbf{g}_n - 40\mathbf{g}_{n+1} - 3\hat{\mathbf{g}}_{n+2}\}, \tag{3.19a}$$

where $\hat{\mathbf{f}}_{n+2} = \mathbf{f}(x_{n+2}, \hat{\mathbf{y}}_{n+2})$, $\hat{\mathbf{g}}_{n+2} = \mathbf{g}(x_{n+2}, \hat{\mathbf{y}}_{n+2})$. It is assumed that the required solution at the step point x_n is known and that we wish to calculate the solution at the point x_{n+1}. In order to calculate the approximation $\hat{\mathbf{y}}_{n+2}$, use is made of the additional scheme

$$\hat{\mathbf{y}}_{n+2} - 32\mathbf{y}_{n+1} + 31\mathbf{y}_n = -h\{14\mathbf{f}_n + 16\mathbf{f}_{n+1}\} + h^2\{-2\mathbf{g}_n + 4\mathbf{g}_{n+1}\}. \tag{3.19b}$$

The equations (3.19a) and (3.19b) may be solved simultaneously for the values $\mathbf{y}_{n+1}$ and $\hat{\mathbf{y}}_{n+2}$ by the use of a standard iterative technique with the initial approximation

$$\mathbf{y}_1 = \mathbf{y}_0 + h\mathbf{f}_0 + \frac{h^2}{2}\mathbf{g}_0.$$

The converged value of $\mathbf{y}_{n+1}$ found by the combination of schemes (3.19a) and (3.19b) is taken as the finally accepted approximation at the point x_{n+1}, but $\hat{\mathbf{y}}_{n+2}$ is not accepted as a final approximation at x_{n+2} on account of its generally inferior accuracy. Having calculated $\mathbf{y}_{n+1}$ we may replace n by $n+1$ in (3.19a, b) and then carry out the same procedure to calculate $\mathbf{y}_{n+2}$ where the initial approximation $\mathbf{y}_{n+2} = \hat{\mathbf{y}}_{n+2}$ is used. It is clear that this approach has some similarities with that outlined earlier in this section, since $\mathbf{y}_{n+1}$ is calculated in terms of the known solution $\mathbf{y}_n$ as well as the unknown solution at an advanced steppoint x_{n+2}. Urabe has proved the convergence of his process for sufficiently small h and the whole procedure seems to be quite useful for the numerical integration of non-stiff systems in cases where the second derivative is fairly cheap to evaluate. Practical experience does seem to indicate, however, that this approach usually requires too many function evaluations for it to be competitive with Adams methods used for example in PECE mode. Also these schemes do not seem to possess sufficiently large regions of absolute stability for them to be of any use in integrating stiff systems of equations and so we shall not consider this approach any further.

To conclude this section we present a particular numerical example which serves to illustrate the rate of convergence that we might expect when Scheme (3.18) is applied to a non-stiff problem and, more important, shows the tremendous gain in efficiency that can be had with these methods if an accurate predictor is used. The problem which we consider is the integration of the scalar equation

$$y' = 6x^2y^{1/2} \qquad y(0) = 1$$

using (3.18) with a constant steplength $h = 0{\cdot}01$. In order to use Scheme (3.18) we need one extra initial condition to be available and for the purposes of our example we use the exact value $y_1 = (1 + h^3)^2$. In Table 3.4 we present three sets of results. These arise from an examination of the cases:

(i) No explicit predictor is used and so the initial approximations are $\{y_n^{(0)}\} = 0$ for all n.

(ii) The 'predictor' $y_n^{(0)} \equiv y_1$ is used for all n.

(iii) The third order predictor $y_{n+1}^{(0)} + 4y_n^{(1)} - 5y_{n-1}^{(2)} = h\{4f_n^{(1)} + 2f_{n-1}^{(2)}\}$ is used to generate the sequence $\{y_n^{(0)}\}$.

In order to generate the required solution, iteration scheme (2.37b) was used in

TABLE 3.4

x	Scheme	Solution obtained	True solution	Number of iterations
0·01	(1)	1·000015709	1·000016000	13
	(2)	1·000015662		5
	(3)	1·000016000		2
0·02	(1)	1·000053659	1·000054001	14
	(2)	1·000053552		6
	(3)	1·000054001		2
0·03	(1)	1·000127442	1·000128004	14
	(2)	1·000127194		6
	(3)	1·000128004		2
0·04	(1)	1·000249205	1·000250016	14
	(2)	1·000249120		7
	(3)	1·000250017		2

each case and iterates were calculated until two successive ones differed by less than 10^{-6}. In Table 3.4 we give the results obtained using approaches (i), (ii) and (iii) at the first four steps and compare the final solution with the true solution. As can be seen there is a tremendous increase in computational efficiency if an explicit predictor is used. This conclusion can generally be drawn when schemes of this type are used.

It should be mentioned at this stage that convergence gets progressively slower, since, as we integrate further and further, gradually more iterations are required per step. As we shall see in Section 3.5, however, this problem can generally be overcome. Further it seems advisable that as accurate a predictor as possible should be used with all these schemes.

3.4 Stability aspects of iterative integration procedures

In this section we again consider the approximate numerical solution of (3.1) using the linear multistep method (3.2). In order to generate the required solution of (3.2) we use a slightly more general class of iterative schemes than was considered in the previous section. The class of schemes which we consider is given by

$$\sum_{j=0}^{m}[\alpha_j \mathbf{y}_{n+j}^{(t)} - h\beta_j \mathbf{f}(x_{n+j}, \mathbf{y}_{n+j}^{(t)}) - h^2\gamma_j \mathbf{g}(x_{n+j}, \mathbf{y}_{n+j}^{(t)})]$$

$$+ \sum_{j=m+1}^{k} [\alpha_j \mathbf{y}_{n+j}^{(t-1)} - h\beta_j \mathbf{f}(x_{n+j}, \mathbf{y}_{n+j}^{(t-1)}) - h^2\gamma_j \mathbf{g}(x_{n+j}, \mathbf{y}_{n+j}^{(t-1)})] = 0 \qquad (3.20)$$

and we shall investigate which stability restrictions we need to impose in order that this scheme should be of some practical use for the numerical integration of (3.1). In order to normalise our iteration scheme we find it convenient to set $\alpha_m = -1$ since this simplifies the notation. The first problem which faces us is that of investigating the convergence of (3.20) to a solution of (3.2) starting from any bounded sequence of initial approximations $\{\mathbf{y}_n^{(0)}\}$. In Section 3.1 we proved a theorem giving sufficient conditions for a particular iterative scheme of the form (3.20) to converge to the solution of the original differential equation. During the proof of this theorem a bound was obtained for the accumulated truncation error committed at any stage. Strict error bounds of this type are notoriously pessimistic and practical experience has shown that convergence may usually be obtained under weaker restrictions than those imposed by Theorem 3.1. In what follows we shall derive a somewhat weaker condition for convergence and as we shall see this leads us to give a definition which extends that of zero-stability.

If we now apply iterative scheme (3.20) to the solution of the scalar test equation $y' = \lambda y$, where λ is a complex constant with negative real part, and if we define $q = h\lambda$ we obtain

$$\sum_{j=0}^{m} (\alpha_j - q\beta_j - q^2\gamma_j) y_{n+j}^{(t)} + \sum_{j=m+1}^{k} (\alpha_j - q\beta_j - q^2\gamma_j) y_{n+j}^{(t-1)} = 0 \tag{3.21}$$

which we rewrite in the form

$$\sum_{j=0}^{m} \hat{\alpha}_j y_{n+j}^{(t)} + \sum_{j=m+1}^{k} \hat{\alpha}_j y_{n+j}^{(t-1)} = 0. \tag{3.22}$$

As usual we shall assume that we are given m initial values of the form $y(x_j) = y_j$, $j = 0(1)m - 1$, and we shall take as our initial conditions $y_j^{(t)} = y_j$ for all integers $t \geq 0$ and $j = 0(1)m - 1$. In order to derive our criterion for the convergence of (3.22) we rewrite this iteration scheme in the matrix form

$$-\hat{\alpha}_m \begin{bmatrix} y_m^{(t)} \\ y_{m+1}^{(t)} \\ \cdot \\ \cdot \\ \cdot \\ \cdot \\ y_{N+k}^{(t)} \end{bmatrix} = \begin{bmatrix} \hat{\alpha}_0 & \hat{\alpha}_1 & \cdot & \hat{\alpha}_{m-1} & & 0 \\ & \hat{\alpha}_0 & & & \hat{\alpha}_{m-1} & \\ & & \cdot & & & \cdot \\ & & & \cdot & & \cdot \\ & & & & \cdot & \\ & & & & & \hat{\alpha}_0 \\ & & & & & \\ & & & & & 0 \end{bmatrix} \begin{bmatrix} y_0 \\ y_1 \\ \cdot \\ \cdot \\ y_{m-1} \\ 0 \\ \cdot \\ 0 \end{bmatrix}$$

$$+ \begin{bmatrix} 0 & & & & & & \\ \hat{\alpha}_{m-1} & 0 & & & & & \\ \hat{\alpha}_{m-2} & \hat{\alpha}_{m-1} & 0 & & & & \\ \cdot & & & & & & \\ \hat{\alpha}_0 & \hat{\alpha}_1 & \cdot & \cdot & \hat{\alpha}_{m-1} & 0 & \\ & \cdot & \cdot & & & & \\ & & \hat{\alpha}_0 & \hat{\alpha}_1 & \cdot & \hat{\alpha}_{m-1} & 0 \end{bmatrix} \begin{bmatrix} y_m^{(t)} \\ y_{m+1}^{(t)} \\ \cdot \\ \cdot \\ \cdot \\ y_{N+m}^{(t)} \\ \cdot \\ y_{N+k}^{(t)} \end{bmatrix}$$

$$
+\begin{bmatrix} 0 & \hat{\alpha}_{m+1} & \hat{\alpha}_{m+2} & \cdot & \hat{\alpha}_k & & & & \\ & 0 & \hat{\alpha}_{m+1} & & \cdot & \hat{\alpha}_k & & & \\ & & \cdot & \cdot & & & \cdot & & \\ & & & 0 & \hat{\alpha}_{m+1} & & \cdot & \hat{\alpha}_k & \\ & & & & 1 & & & & \\ & & & & & 1 & \cdot & 0 & \\ & & & & & & & \cdot & 1 \end{bmatrix} \begin{bmatrix} y_m^{(t-1)} \\ y_{m+1}^{(t-1)} \\ \cdot \\ \cdot \\ y_{N+m}^{(t-1)} \\ \cdot \\ \cdot \\ \cdot \\ y_{N+k}^{(t-1)} \end{bmatrix} \tag{3.23}
$$

where N is some large integer with magnitude depending on the range over which the solution of (3.20) is required. In order to keep the dimension of our system constant we define $y_i^{(t)} = y_i^{(t-1)}$ for all $i \geq N + m + 1$. This in turn implies that if convergence in some appropriate sense is obtained in t iterations then only the iterates $y_m^{(t)}, y_{m+1}^{(t)}, \ldots, y_{N+k+t(m-k)}^{(t)}$ can be guaranteed to be solutions of our recurrence relation. We note that this implies that N must be chosen to be sufficiently large so that if convergence is required in the range $m \leq i \leq L$ then $N + k + t(m - k) \geq L$. As we shall see later this does not present any difficulty in practice. Our procedure for defining the iterates at the end of the range of integration ($i > N + k + t(m - k)$) will not for two reasons affect our analysis. Firstly in examining the stability properties of our scheme we shall restrict t to be finite and secondly we shall be interested in the behaviour of our solution only for very large N. Note that all the matrices appearing in relation (3.23) have well-defined sparseness structures. It is convenient to rewrite this system of equations in the form

$$
y^{(t)} = A_1 y_0 + A_2 y^{(t)} + A_3 y^{(t-1)},
$$

where A_1, A_2, and A_3 are all of size $(N + k - m + 1) \times (N + k - m + 1)$. Rearranging this system of equations we have

$$
(I - A_2) y^{(t)} = A_1 y_0 + A_3 y^{(t-1)}
$$

or

$$
y^{(t)} = A^{-1} B y^{(t-1)} + A^{-1} C y_0, \tag{3.24}
$$

where $A = I - A_2$, $B = A_3$, $C = A_1$. It now follows immediately that this iteration scheme converges for any bounded sequence of initial approximations $y^{(0)}$ if

$$\rho(M) < 1, \tag{3.25}$$

where $M = A^{-1}B$. The inequality (3.25) illustrates the very close connection between (3.23) and the Gauss–Seidel iteration scheme. In fact, as $N \to \infty$ our scheme (3.23) becomes identical with Gauss–Seidel iteration and this fact allows us to use the following sufficient condition for convergence, the proof of which may be found in Isaacson and Keller (1966), p. 67.

Theorem 3.2

Let

$$r \equiv \sum_{\substack{j=0 \\ j \neq m}}^{k} \frac{|\hat{\alpha}_j|}{|\hat{\alpha}_m|},$$

where $r < 1$. *Define an error vector* $\mathbf{e}^{(t)} = \mathbf{y}^{(t)} - \mathbf{y}$. *Then* $\|\mathbf{e}^{(t)}\|_\infty \leq r^t \|\mathbf{e}^{(0)}\|_\infty$ *and so* $\|\mathbf{e}^{(t)}\|_\infty \to 0$ *as* $t \to \infty$ *ensuring the convergence of our iteration scheme* (3.23).

If we let $q = 0$ in (3.21) we obtain the recurrence relation

$$\sum_{j=0}^{m} \alpha_j y_{n+j}^{(t)} + \sum_{j=m+1}^{k} \alpha_j y_{n+j}^{(t-1)} = 0. \tag{3.26}$$

We now make the important simplifying assumption that $\alpha_i \geq 0$, $0 \leq i \leq k$, $i \neq m$, since this is the only class of schemes which will be considered in this chapter. Because of the consistency of our scheme (3.20), this gives rise to the case $r = 1$ which is of little practical interest, because if the resulting scheme is convergent it is too slowly convergent to be of any practical use. The case $r = 1$ is, however, of theoretical interest since it allows us to extend the concept of zero-stability. To prove convergence of (3.20) we recall at this stage one of the important assumptions of Theorem 3.1 which was that the initial approximations $\{y_n^{(0)}\}$ are computed using a separate explicit predictor which is zero-stable.

It follows from this that in the limit $h = 0$ we have that $\{y_n^{(0)}\} \equiv y_n$ where y_n is the required solution of (3.26). It is now trivial to show that the solution $y_n^{(1)}$, which is obtained after one application of (3.26) and which is the finally accepted approximation to y_n since $|y_n^{(0)} - y_n^{(1)}|$ is less than any reasonable prescribed tolerance, satisfies $y_n^{(1)} \equiv y_n$.

Having obtained sufficient conditions for the convergence of the iteration scheme (3.20) it now remains to investigate its stability properties. Applying

(3.2) to the scalar test equation $y' = \lambda y$ we obtain

$$\sum_{j=0}^{k} (\alpha_j - q\beta_j - q^2\gamma_j)y_{n+j} = 0.$$

The general solution of this recurrence relation may be written in the form

$$y_n = \sum_{i=1}^{k} c_i b_i^n, \tag{3.27}$$

where the c_i's are constants and the b_i's are the k roots, assumed distinct, of the polynomial

$$\sum_{j=0}^{k} (\alpha_j - q\beta_j - q^2\gamma_j)x^j = 0.$$

If the iteration scheme (3.20) converges to the prescribed degree of precision in t iterations, it follows that we may express $y_n^{(t)}$ in the form

$$y_n^{(t)} = \sum_{i=1}^{k} c_i^{(n)} b_i^n + \varepsilon_n, \tag{3.28}$$

where ε_n is a local error which has magnitude depending on the degree of precision to which the computations are performed. It is usually the case that some of the roots b_i have modulus greater than 1 and we shall assume that there are s roots, numbered from b_1 to b_s, with this property. In order that the scheme (3.20) should be stable it is clearly necessary for

$$c_1^{(n)} = c_2^{(n)} = \cdots = c_{(s)}^{(n)} = 0, \ \forall \ n.$$

If any one of these coefficients is non-zero the solution $y_n^{(t)}$ will increase without bound. In view of this, when establishing the stability of the scheme (3.20), it is sufficient to show that the sequence $\{y_n^{(t)}\}$ remains bounded for all n and fixed t. This is the approach which we shall adopt in our subsequent analysis. A sufficient condition for the boundedness of our solution is contained in the following theorem.

Theorem 3.3

Consider the approximate numerical solution of the scalar linear recurrence relation

$$\sum_{j=0}^{k} \hat{\alpha}_j y_{n+j} = 0,$$

using the iteration scheme (3.22). Suppose now that we use the initial conditions

$$y_j^{(t)} = y(x_j) \equiv y_j; \qquad j = 0, 1, \ldots, m-1, t \geq 0,$$

where $|y_j| \leq W < \infty$, *and initial approximations*

$\{y_n^{(0)}\}, \qquad n \geq m$ *with* $|y_n^{(0)}| \leq Y^{(0)} < \infty$ *for all n.*

Let

$$\alpha = \sum_{j=0}^{m-1} \frac{|\hat{\alpha}_j|}{|\hat{\alpha}_m|} \quad \text{and} \quad \beta = \sum_{j=m+1}^{k} \frac{|\hat{\alpha}_j|}{|\hat{\alpha}_m|};$$

then providing that

$$(1) \quad \alpha + \beta < 1$$

the sequence $\{y_n^{(t)}\}$ *generated by (3.22) is bounded for all n and finite t in the sense that* $|y_n^{(t)}| \leq Y^{(t)} < \infty$ *for some sequence of positive constants* $Y^{(t)}$.

Proof

The proof of this Theorem will be by induction. We shall first of all establish the relation

$$|y_{n+m}^{(1)}| \leq \frac{\alpha}{1-\alpha} W + \frac{\beta}{1-\alpha} Y^{(0)}. \tag{3.28a}$$

Substituting the values $n = 0$, $t = 1$ into (3.22) we have

$$|y_m^{(1)}| \leq \alpha W + \beta Y^{(0)}.$$

This establishes the validity of (3.28a) for $n = 0$. It is also trivial to show that

$$|y_j^{(1)}| \leq \frac{\alpha}{1-\alpha} W + \frac{\beta}{1-\alpha} Y^{(0)} \quad \text{for all} \quad m \leq j \leq 2m-1.$$

Assuming that (3.28a) is valid for $n = l$ it follows that

$$|y_{l+m}^{(1)}| \leq \frac{\alpha}{1-\alpha} W + \frac{\beta}{1-\alpha} Y^{(0)}.$$

Substituting into (3.22) for $t = 1$, $n = l + 1$ we have

$$|y_{m+l+1}^{(1)}| \leq \frac{\alpha^2}{1-\alpha} W + \frac{\alpha\beta}{1-\alpha} Y^{(0)} + \beta Y^{(0)}$$

$$\leq \frac{\alpha}{1-\alpha} W + \frac{\beta Y^{(0)}}{1-\alpha}.$$

Hence the validity of (3.28a) is established by induction. It now follows that

$$\max_n |y_n^{(1)}| \equiv Y^{(1)} \leq \frac{\beta Y^{(0)}}{1-\alpha} + \frac{\alpha}{1-\alpha} W.$$

Applying exactly the same argument for general t we have

$$\begin{aligned}
\max_n |y_n^{(t)}| \equiv Y^{(t)} &\leq \frac{\beta Y^{(t-1)}}{1-\alpha} + \frac{\alpha}{1-\alpha} W \\
&\leq \frac{\beta}{1-\alpha}\left\{\frac{\beta}{1-\alpha} Y^{(t-2)} + \frac{\alpha}{1-\alpha} W\right\} + \frac{\alpha}{1-\alpha} W \\
&\cdots\cdots\cdots\cdots\cdots\cdots\cdots\cdots\cdots\cdots \\
&\leq \left(\frac{\beta}{1-\alpha}\right)^t Y^{(0)} + \frac{\alpha}{1-\alpha} W \left(\frac{1}{1 - \dfrac{\beta}{1-\alpha}}\right) \\
&\leq \left(\frac{\beta}{1-\alpha}\right)^t Y^{(0)} + \frac{\alpha}{1-\alpha-\beta} W.
\end{aligned}$$

From condition (1) of Theorem 3.3 it follows that $y_n^{(t)}$ is bounded for all n and t.

On account of this analysis it is natural to give the following definition of *iterative absolute stability*.

Definition 3.2

The iteration scheme (3.20) is said to possess the property of iterative absolute stability for any fixed value of $q = h\lambda$ if, for that value of $h\lambda$ and for any bounded sequence of initial approximations $\{y_n^{(0)}\}$,

(i) the sequence $\{y_n^{(t)}\}$ generated from (3.21) is convergent in the sense that, given any error tolerance ε, there exists a T such that

$$|y_n^{(t)} - y_n^{(t+1)}| < \varepsilon \text{ for all } t \geq T \text{ and all } n = m, m+1, \ldots$$

(ii) $$\lim_{\substack{x \to \infty \\ x = nh + x_0}} y_n^{(t)} = 0 \text{ for all } t \geq T,$$

where we assume that t is finite.

A region R of the complex plane is said to be a *region of iterative absolute stability* of (3.20) if this equation possesses the property of iterative absolute stability for all $q \in R$.

Definition 3.3

The iteration scheme (3.20) is said to be *iteratively A-stable* if its region of iterative absolute stability contains the whole of the complex left hand half plane $\text{Re}(q) < 0$.

Practical experience seems to indicate that it is still necessary for an iteratively A-stable scheme based on a linear multistep method to be implicit. However, as we shall show later on in this chapter, it is possible to develop iteratively A-stable schemes of order greater than 2. We may also define the concept of *iterative zero-stability* in the following way.

Definition 3.4

The iterative scheme (3.20) is said to be iteratively zero-stable if there exists a region $B_R \equiv \{q{:}0 < |q| < \hat{\varepsilon}, \text{Re}(q) < 0\}$ such that (3.20) possesses the property of iterative absolute stability for all $q \in B_R$.

We note that this definition is not strictly a direct extension of zero-stability since an iteratively zero-stable scheme must possess the property of iterative absolute stability for some sufficiently small $q < 0$, whereas a zero-stable scheme need not have any region of absolute stability.

We may now summarise the results contained in Theorems 3.2 and 3.3 obtained earlier in this section with the following lemma.

Lemma 3.1

Consider the numerical solution of the scalar test equation $y' = \lambda y$ using iteration scheme (3.20) *with a fixed value of h. Suppose that the resulting recurrence relation* (3.22) *is such that $\alpha + \beta < 1$ where*

$$\alpha = \sum_{j=0}^{m-1} \frac{|\hat{\alpha}_j|}{|\hat{\alpha}_m|} \quad \textit{and} \quad \beta = \sum_{j=m+1}^{k} \frac{|\hat{\alpha}_j|}{|\hat{\alpha}_m|}.$$

Then (3.20) *possesses the property of iterative absolute stability for the particular value $q = h\lambda$.*

The proof of this lemma is not given explicitly but follows from a combination of Theorems 3.2 and 3.3. We now give an example of the practical applications of this lemma. This consists of finding a sufficient condition for the point $q = h\lambda$ to lie inside the region of iterative absolute stability of (3.18). Applying scheme (3.18) to the scalar test equation $y' = \lambda y$ where, in order to simplify our analysis, we shall assume that λ is real and negative we obtain

$$y_{n+3}^{(t-1)} - 6y_{n+2}^{(t)} + (3 + 6q)y_{n+1}^{(t)} + 2y_n^{(t)} = 0. \tag{3.29}$$

Using Lemma 3.1 it follows that a sufficient condition for the point q to lie inside the interval of iterative absolute stability of (3.18) is for

$$\frac{1}{6} + \frac{2}{6} + \left|\frac{3 + 6q}{6}\right| < 1, \quad \text{i.e.} \quad q \in (-1, 0).$$

We shall not consider the implications of our definitions any more in this section. Instead we proceed to consider some algorithms of practical interest in the next section.

3.5 Stiff systems of equations

The algorithms developed in Section 3.1 are in general inappropriate for the numerical integration of stiff systems of ordinary differential equations since the steplength h has to be taken excessively small in order that the convergence of our secondary iteration scheme may be guaranteed. This arises since, in regions where (3.1) is stiff, the Lipschitz constants K_1 and K_2 are normally very large. Stiff systems are in fact sometimes referred to as systems with large Lipschitz constants. The inadequacy of the algorithms of Section 3.1 therefore makes it necessary that more efficient schemes for the integration of stiff systems be derived. These are based on the use of a more sophisticated implicit secondary iteration scheme because any explicit secondary scheme, such as a direct iteration, will not converge with a value of h appropriate for sampling the required slowly varying solution. This approach will be discussed in Section 3.6.

In order to simplify our notation in the forthcoming sections, it is assumed that the explicit dependence on x appearing on the right hand side of (3.1) has been removed in the usual way through the introduction of an additional variable $\hat{y}$ which is such that $\hat{y}' = 1$ with $\hat{y}(0) = 0$. The original system of differential equations may thus be assumed to be in the form $\mathbf{y}' = \mathbf{f}(\mathbf{y})$ where $\mathbf{y}$ is an s-vector.

3.6 A high order one-step integration formula

The basic ideas in this section are explained in the context of linear two-step schemes. Consider first the primary iteration scheme

$$\alpha_2 y_{n+2}^{(t-1)} + \alpha_1 y_{n+1}^{(t)} + \alpha_0 y_n^{(t)}$$
$$= h\{\beta_2 f(y_{n+2}^{(t-1)}) + \beta_1 f(y_{n+1}^{(t)}) + \beta_0 f(y_n^{(t)})\}$$
$$+ h^2\{\gamma_2 g(y_{n+2}^{(t-1)}) + \gamma_1 g(y_{n+1}^{(t)}) + \gamma_0 g(y_n^{(t)})\} \qquad (3.30)$$

which corresponds to the solution of a "tri-diagonal" system of non-linear algebraic equations using a primary iteration scheme of the Gauss–Seidel type. In an attempt to develop an algorithm which is efficient for the numerical solution of stiff systems of equations, we shall generate the required solution, $y_{n+1}^{(t)}$, of (3.30) using a modified Newton iteration scheme. When designing particular integration procedures we shall attempt to develop algorithms which are such that the primary iteration scheme converges rapidly for linear systems, in the hope that this desirable property will also be carried over to non-linear systems. Suppose then that we apply (3.30) to the solution of the scalar test equation $y' = \lambda y$, where λ is a complex constant with negative real part. This produces the relation

$$y_{n+1}^{(t)} = \left[\frac{-\alpha_0 + h\lambda\beta_0 + h^2\lambda^2\gamma_0}{\alpha_1 - h\lambda\beta_1 - h^2\lambda^2\gamma_1}\right] y_n^{(t)} + \left[\frac{-\alpha_2 + h\lambda\beta_2 + h^2\lambda^2\gamma_2}{\alpha_1 - h\lambda\beta_1 - h^2\lambda^2\gamma_1}\right] y_{n+2}^{(t-1)}. \quad (3.31)$$

In general the main contribution to the error in $y_{n+1}^{(t)}$ will come from the error in the term $y_{n+2}^{(t-1)}$ and so it would clearly be advantageous if we could make the coefficient of $y_{n+2}^{(t-1)}$ small for all values of $q = h\lambda$ with $\text{Re}(q) < 0$. This aim can be achieved for $|q|$ small if we set $\alpha_2 = 0$ (and then to normalise our scheme we set $\alpha_1 = 1$ and for consistency $\alpha_0 = -1$) and for $|q|$ large if we set $\gamma_2 = 0$, $\gamma_1 \neq 0$. By constraining the coefficient of $y_{n+2}^{(t-1)}$ to be small for both very small and very large values of $|q|$ we may hope to derive schemes which are such that this coefficient remains small for all $|q|$ with $\text{Re}\,(q) < 0$. This would seem in general to be a necessary condition for our primary iteration scheme to converge rapidly. In view of this we shall from now on be concerned only with schemes of the form (3.30) for which $\alpha_2 = \gamma_2 = 0$.

A scheme of this particular form, which has been found to be useful for the numerical integration of stiff systems, is based on the fifth order method

$$\mathbf{y}_{n+1} - \mathbf{y}_n = h\{\tfrac{1}{120}\mathbf{f}_{n+2} + \tfrac{8}{15}\mathbf{f}_{n+1} + \tfrac{11}{24}\mathbf{f}_n\}$$
$$+ h^2\{-\tfrac{7}{60}\mathbf{g}_{n+1} + \tfrac{1}{15}\mathbf{g}_n\}. \qquad (3.32)$$

Formula (3.32) is rather like an Adams-type method of the kind discussed in Section 3.2. An expression for the principal term in the local truncation error of (3.32) can be found using the procedure described in Section 3.1 and may be shown to be $-(1/2400)h^6\mathbf{y}_n^{(\text{VI})}$. In order that we may be able to generate the required solution of (3.32) in an efficient manner, we use a primary iteration scheme based on a non-linear Gauss–Seidel iteration and a secondary iteration scheme based on a modified Newton scheme resulting from the substitutions

$$\left.\begin{aligned} \mathbf{f}_n^{(t)} &= \mathbf{f}_n^{(t-1)} + \mathbf{J}_n^{(t-1)}\{\mathbf{y}_n^{(t)} - \mathbf{y}_n^{(t-1)}\} \\ &\text{and} \\ \mathbf{g}_n^{(t)} &\equiv \mathbf{J}_n^{(t)}\mathbf{f}_n^{(t)}, \\ &= \mathbf{J}_n^{(t-1)}\{\mathbf{f}_n^{(t-1)} + \mathbf{J}_n^{(t-1)}[\mathbf{y}_n^{(t)} - \mathbf{y}_n^{(t-1)}]\}, \end{aligned}\right\} \tag{3.33}$$

where

$$\mathbf{f}_n^{(t)} = \mathbf{f}(\mathbf{y}_n^{(t)}) \quad \text{and} \quad \mathbf{J}_n^{(t)} = \frac{\partial \mathbf{f}(y)}{\partial \mathbf{y}}\bigg|\,\mathbf{y} = \mathbf{y}_n^{(t)}.$$

Putting $\mathbf{D}_n^{(t)} = \mathbf{y}_n^{(t)} - \mathbf{y}_n^{(t-1)}$ and substituting (3.33) into (3.32) with (3.30) used as the primary iteration scheme we obtain the one-step Gauss–Seidel–modified Newton iteration scheme

$$\begin{aligned} &\{\mathbf{I} - \tfrac{8}{15}h\mathbf{J}_{n+1}^{(t-1)} + \tfrac{7}{60}h^2[\mathbf{J}_{n+1}^{(t-1)}]^2\}\mathbf{D}_{n+1}^{(t)} \\ &\quad = -\mathbf{y}_{n+1}^{(t-1)} + \mathbf{y}_n^{(t)} + h\{\tfrac{1}{120}\mathbf{f}_{n+2}^{(t-1)} + \tfrac{8}{15}\mathbf{f}_{n+1}^{(t-1)} + \tfrac{11}{24}\mathbf{f}_n^{(t)}\} \\ &\qquad + h^2\{-\tfrac{7}{60}\mathbf{J}_{n+1}^{(t-1)}\mathbf{f}_{n+1}^{(t-1)} + \tfrac{1}{15}\mathbf{J}_n^{(t)}\mathbf{f}_n^{(t)}\} \end{aligned} \tag{3.34}$$

which now depends only on the single iteration parameter t. This scheme will clearly converge to the required solution for small enough h since in the limit as $h \to 0$ it becomes $\mathbf{y}_n^{(t)} = \mathbf{y}_{n+1}^{(t)}$. If this scheme does converge then it is clear that it does so to a solution of the original equation (3.32). Practical experience which has so far been gained has shown that stiffness does not impose a severe limitation on the stepsize of integration which can be used with scheme (3.34). Rapid convergence to the required solution is usually obtained with a reasonably large value of h providing $h\lambda_i \in R$ for all $i = 1, 2, \ldots, s$, where λ_i are the localised eigenvalues of the linearised problem (these of course vary over the range of integration), and where R is the region of iterative absolute stability of (3.34). It is also necessary that we supply reasonably accurate initial approximations for use in our modified Newton iteration scheme. In order to keep the amount of storage space required to a minimum and to allow changes in the steplength of integration to be performed in a relatively efficient manner, we calculate the sequence of iterates in the order defined by Fig. 2.1.

We now consider a simple procedure for finding the approximate region of iterative absolute stability of scheme (3.34). If we apply this scheme to the scalar test equation $y' = \lambda y$ we obtain a relation of the form (3.31) with appropriate values of the constants α_i, β_i and γ_i, i.e.

$$y_{n+1}^{(t)} = \left[\frac{1 + \frac{11}{24}q + \frac{1}{15}q^2}{1 - \frac{8}{15}q + \frac{7}{60}q^2}\right] y_n^{(t)} + \left[\frac{\frac{1}{120}q}{1 - \frac{8}{15}q + \frac{7}{60}q^2}\right] y_{n+2}^{(t-1)}. \tag{3.35}$$

This iteration procedure is precisely the Gauss–Seidel scheme for the solution of the tri-diagonal system of linear algebraic equations which arises from applying (3.34) to our scalar test equation. We may now compute an approximate region of iterative absolute stability of this scheme by applying the analysis developed in Section 3.4 and in particular by using the condition given by **Lemma 3.1**. For our particular problem this sufficient condition reduces to the requirement that

$$\left|\frac{1 + \frac{11}{24}q + \frac{1}{15}q^2}{1 - \frac{8}{15}q + \frac{7}{60}q^2}\right| + \left|\frac{\frac{1}{120}q}{1 - \frac{8}{15}q + \frac{7}{60}q^2}\right| < 1. \tag{3.36}$$

The region G where inequality (3.36) holds is shown in Fig. 3.3. As mentioned previously it follows that if R is the true region of iterative absolute stability of our scheme then $G \subseteq R$. There would not in general seem to be any convenient numerical procedure for determining the region R and so in practice the best that we can do is to plot G. It can be seen from Fig. 3.3 that scheme (3.34) has a

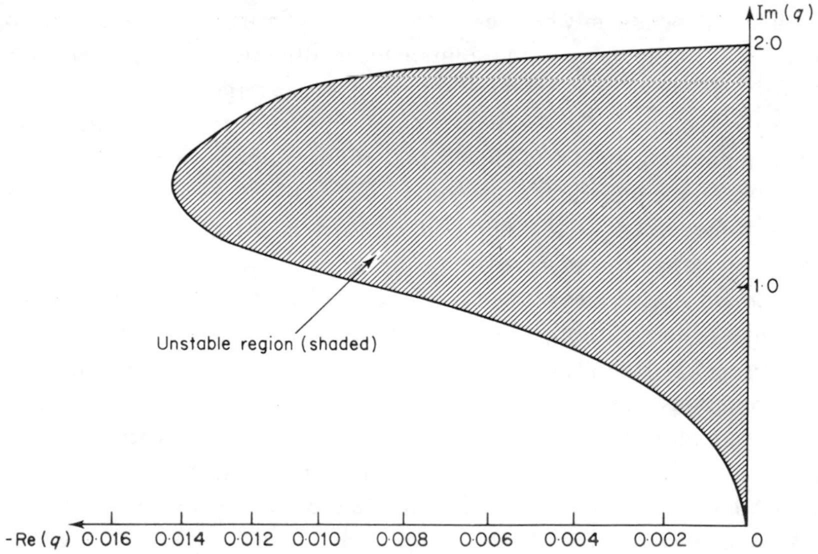

FIG. 3.3 Approximate region, G, of absolute stability of (3.34).

very large region of iterative absolute stability and in particular is (iteratively) $A(0)$-stable. In practice, when solving non-linear systems of ordinary differential equations, it is not normally necessary to know the exact region of iterative absolute stability of our scheme since, as the eigenvalues λ_i are varying with y, it is desirable that at each step the points $h\lambda_i$ should lie well inside the region R. This criterion is usually satisfied if all points lie inside G. Practical experience has also shown that G is usually a "good approximation" to R. It is very important to realise at this stage that, for a particular problem and a particular sequence of initial approximations $\{y_n^{(0)}\}$, perfectly satisfactory results may often be obtained with a value of h which is such that $h\lambda_i \notin R$ for at least one of the eigenvalues λ_i. In view of this the implications of our definition of iterative absolute stability are somewhat different from those which arise from the more usual definition of absolute stability. The main reason for this is that our definition applies for each and every bounded sequence of initial approximations $\{y_n^{(0)}\}$. It is clear, however, that for any given particular problem the region for which we obtain convergence to an ultimately decreasing solution of this problem may be larger than the region of iterative absolute stability of our integration method.

Experience would also seem to indicate that approximations (3.33), which are specially designed so that the matrix $(\partial^2 \mathbf{f}/\partial \mathbf{y}^2)$ does not need to be computed, do not inhibit convergence and good results can usually be obtained with a reasonably large value of h. Note that it would be necessary to compute this matrix if a straightforward Newton iteration scheme were to be used as the secondary iteration. Good results using these approximations have also been reported by Liniger and Willoughby (1967). For numerical examples the reader is referred to Cash (1978a).

3.7 Extension to the multistep case

Before examining the practical aspects of the class of iterative algorithms defined by (3.30) in any great detail, we shall first of all examine their extension to schemes with stepnumber greater than two. The basic ideas which we are proposing can best be explained by considering the three-step case

$$\sum_{j=0}^{3} \alpha_j \mathbf{y}_{n+j} = h \sum_{j=0}^{3} \beta_j \mathbf{f}_{n+j} + h^2 \sum_{j=0}^{3} \gamma_j \mathbf{g}_{n+j}, \tag{3.37}$$

since the extension to even higher order cases is relatively straightforward once this case is well understood. Following the procedure adopted earlier we shall again develop schemes for which the primary iteration converges rapidly for both very small and very large values of $|h\lambda|$ when applied to our scalar test equation $y' = \lambda y$. The hope is that an algorithm will be found for which the primary iteration scheme converges rapidly for all $h\lambda$ with $\text{Re}(h\lambda) < 0$. We may then hope that, if this aim is achieved, this property of rapid convergence of our primary iteration scheme will carry over to the non-linear case as well. If we attempt to generate a solution of (3.37) using an iterative scheme of the form (3.20) the possible choices of the integer m are either 1 or 2. Practical experience has shown it to be difficult to develop efficient algorithms if we choose $m = 1$ and so for the solution of (3.37) we shall confine our attention to primary iteration schemes of the form

$$\sum_{j=0}^{2} [\alpha_j \mathbf{y}_{n+j}^{(t)} - h\beta_j \mathbf{f}_{n+j}^{(t)} - h^2\gamma_j \mathbf{g}_{n+j}^{(t)}] = -\alpha_3 \mathbf{y}_{n+3}^{(t-1)} + h\beta_3 \mathbf{f}_{n+3}^{(t-1)} + h^2\gamma_3 \mathbf{g}_{n+3}^{(t-1)}, \tag{3.38}$$

where $\mathbf{f}_n^{(t)} = \mathbf{f}(\mathbf{y}_n^{(t)})$, etc. Applying (3.38) to the scalar equation $y' = \lambda y$ we obtain

$$[\alpha_2 - h\lambda\beta_2 - h^2\lambda^2\gamma_2]y_{n+2}^{(t)} + [\alpha_1 - h\lambda\beta_1 - h^2\lambda^2\gamma_1]y_{n+1}^{(t)} + [\alpha_0 - h\lambda\beta_0 - h^2\lambda^2\gamma_0]y_n^{(t)} = [-\alpha_3 + h\lambda\beta_3 + h^2\lambda^2\gamma_3]y_{n+3}^{(t-1)}. \tag{3.39}$$

In general we find that the main contribution to the error in $y_{n+2}^{(t)}$ comes from the error in $y_{n+3}^{(t-1)}$ and this contribution can be made small for both very small and very large values of $|h\lambda|$ by choosing $\alpha_3 = \gamma_3 = 0$. In order to normalise our scheme we now set $\alpha_2 = 1$. Given that $\alpha_3 = \gamma_3 = 0$ the highest attainable order of accuracy of scheme (3.38) is 8. It turns out however that this scheme has only a finite region of absolute stability and so is rejected for the purpose of solving stiff systems. If however we consider the set of seventh order schemes of the form (3.38), all of which contain one free parameter β_3, we find that for a certain range of β_3 the resulting iteration scheme has an infinite region of iterative absolute stability. The general seventh order scheme of the form (3.38) is given by

$$\begin{aligned}\mathbf{y}_{n+2}^{(t)} - \left(\frac{192 - 45360\beta_3}{96}\right)\mathbf{y}_{n+1}^{(t)} + \left(\frac{96 - 45360\beta_3}{96}\right)\mathbf{y}_n^{(t)} \\ = h\left\{\beta_3 \mathbf{f}_{n+3}^{(t-1)} + \left(\frac{12 + 1269\beta_3}{32}\right)\mathbf{f}_{n+2}^{(t)} + \frac{23328}{96}\beta_3 \mathbf{f}_{n+1}^{(t)} \right. \\ \left. + \left(\frac{6043\beta_3 - 12}{32}\right)\mathbf{f}_n^{(t)}\right\} + h^2\left\{\left(\frac{-4 - 1431\beta_3}{96}\right)\mathbf{g}_{n+2}^{(t)} \right. \\ \left. + \left(\frac{64 - 18576\beta_3}{192}\right)\mathbf{g}_{n+1}^{(t)} + \left(\frac{2169\beta_3 - 4}{96}\right)\mathbf{g}_n^{(t)}\right\}.\end{aligned} \tag{3.40}$$

If we now apply this scheme to the scalar test equation $y' = \lambda y$ we obtain

$$-h\beta_3\lambda y_{n+3}^{(t-1)} + \left[1 - \left(\frac{12 + 1269\beta_3}{32}\right)h\lambda + \left(\frac{4 + 1431\beta_3}{96}\right)h^2\lambda^2\right]y_{n+2}^{(t)}$$
$$- \left[\frac{192 - 45360\beta_3}{96} + \frac{23328\beta_3 h\lambda}{96} + \frac{(64 - 18576\beta_3)}{192}h^2\lambda^2\right]y_{n+1}^{(t)}$$
$$+ \left[\frac{96 - 45360\beta_3}{96} - \frac{(6043\beta_3 - 12)}{32}h\lambda - \frac{(2169\beta_3 - 4)}{96}h^2\lambda^2\right]y_n^{(t)}$$
$$= 0. \tag{3.41}$$

This iteration scheme may be written in the form

$$ay_{n+3}^{(t-1)} + by_{n+2}^{(t)} + cy_{n+1}^{(t)} + dy_n^{(t)} = 0 \tag{3.42}$$

for suitable values of the constants a, b, c and d. This is precisely a Gauss–Seidel iteration scheme for the solution of a particular system of linear algebraic equations where the matrices **D**, **L** and **U** usually associated with this scheme are defined by

$$\mathbf{D} = \begin{bmatrix} b & & & & 0 \\ & b & & & \\ & & \cdot & & \\ & & & \cdot & \\ & & & & \cdot \\ 0 & & & & b \end{bmatrix}, \quad -\mathbf{U} = \begin{bmatrix} 0 & a & & & 0 \\ & 0 & a & & \\ & & \cdot & \cdot & \\ & & & \cdot & \cdot \\ & & & 0 & a \\ 0 & & & & 0 \end{bmatrix}, \quad -\mathbf{L} = \begin{bmatrix} 0 & & & & & \\ c & 0 & & & & \\ d & c & 0 & & & \\ & d & c & 0 & & \\ & & \cdot & \cdot & \cdot & \\ & & & \cdot & \cdot & \\ & & & d & c & 0 \end{bmatrix}$$

It follows from Lemma 3.1 that a sufficient condition for iteration scheme (3.42) to converge to a decreasing solution is that

$$\left|\frac{a}{b}\right| + \left|\frac{c}{b}\right| + \left|\frac{d}{b}\right| < 1. \tag{3.43}$$

It now follows immediately that if (3.43) holds for a particular value of $q = h\lambda$ then $q \in R$ where R is the region of iterative absolute stability of (3.40). We define a region G of the complex plane as that set of q which is such that (3.43) is valid and it follows immediately that $G \subseteq R$. By plotting the regions $G(\beta_3)$ for a range of the free parameter β_3 it has been found that G contains the negative real axis (and so scheme (3.40) is iteratively $A(0)$-stable) for all β_3

lying in the range [4/1425, 20/5013] with the value $\beta_3 = 27/8500$ giving rise to one of the largest regions for G. In Fig. 3.4 the region G is plotted and as can be seen scheme (3.40) has a large region of iterative absolute stability.

In order to generate the required solution $y_{n+2}^{(t)}$ of (3.40) in the non-linear case we use a one-step modified Newton iteration scheme as our secondary iteration, where we again make use of approximations (3.33). We may now write our one-step Gauss–Seidel–modified Newton iteration scheme in the form

$$
\begin{aligned}
\{\mathbf{I} - \beta_2 h\mathbf{J}_{n+2}^{(t-1)} - \tfrac{1}{2}\gamma_2 h^2 [\mathbf{J}_{n+2}^{(t-1)}]^2\}\{\mathbf{y}_{n+2}^{(t)} - \mathbf{y}_{n+2}^{(t-1)}\} \\
= -\mathbf{y}_{n+2}^{(t-1)} - \alpha_1 \mathbf{y}_{n+1}^{(t)} - \alpha_0 \mathbf{y}_n^{(t)} \\
+ h\{\beta_3 \mathbf{f}_{n+3}^{(t-1)} + \beta_2 \mathbf{f}_{n+2}^{(t-1)} + \beta_1 \mathbf{f}_{n+1}^{(t)} + \beta_0 \mathbf{f}_n^{(t)}\} \\
+ \tfrac{1}{2}h^2\{\gamma_2 \mathbf{J}_{n+2}^{(t-1)}\mathbf{f}_{n+2}^{(t-1)} + \gamma_1 \mathbf{J}_{n+1}^{(t)}\mathbf{f}_{n+1}^{(t)} + \gamma_0 \mathbf{J}_n^{(t)}\mathbf{f}_n^{(t)}\},
\end{aligned}
\tag{3.44}
$$

where

$$
\beta_3 = \frac{27}{8500}, \quad \alpha_0 = \frac{96 - 45360\beta_3}{96}, \quad \alpha_1 = -\frac{(192 - 45360\beta_3)}{96},
$$

$$
\beta_2 = \frac{12 + 1269\beta_3}{32}, \quad \beta_1 = \frac{23328\beta_3}{96}, \quad \beta_0 = \frac{6043\beta_3 - 12}{32},
$$

$$
\gamma_2 = -\frac{(4 + 1431\beta_3)}{48}, \quad \gamma_1 = \frac{64 - 18576\beta_3}{96}, \quad \gamma_0 = \frac{2169\beta_3 - 4}{48}.
$$

When using iteration schemes of the general form (3.44) in practice one of our main considerations will be to keep the number of function and Jacobian evaluations to a minimum. A considerable saving in computational effort can often be obtained with procedures incorporating a modified Newton iteration scheme if we keep the coefficient matrix piecewise constant (see Willoughby, 1974, p. 11 for example). Following this approach we shall keep the matrix

$$
\mathbf{I} - h\beta_2 \mathbf{J}_{n+2}^{(t-1)} - \tfrac{1}{2}\gamma_2 h^2 [\mathbf{J}_{n+2}^{(t-1)}]^2
$$

constant for as long as possible and re-evaluate it only when the iteration scheme (3.44) fails to converge sufficiently rapidly or when h changes. Since this modification is now well known and widely used we shall not discuss it any further at this stage.

The precise order in which the sequence of iterates is calculated is not vitally important as long as the order of calculation is arranged in such a way that the

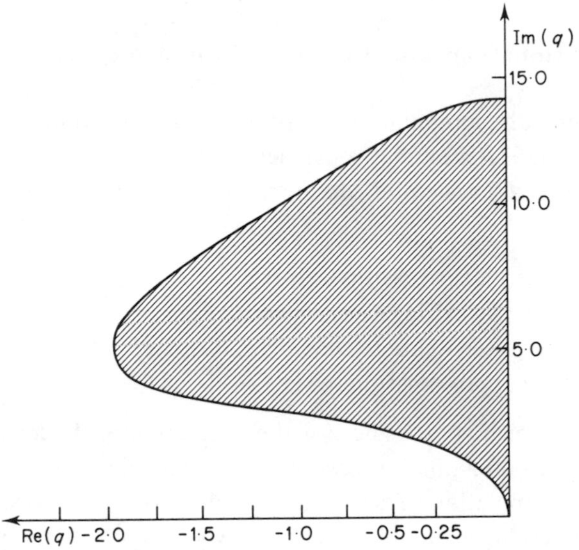

FIG. 3.4 Region of instability (shaded) of (3.40).

solution at x_2 is calculated as quickly as possible without too many iterates at advanced time steps needing to be computed. However, in order to minimise the amount of storage space required, and to allow changes in the steplength of integration to be made in a relatively straightforward fashion, we do the calculation in a precisely defined order. It has been found to be rather inefficient to use the order defined by Fig. 2.1 and so we calculate the required sequence in a slightly different order as defined by Fig. 3.5 which has been found to give satisfactory results.

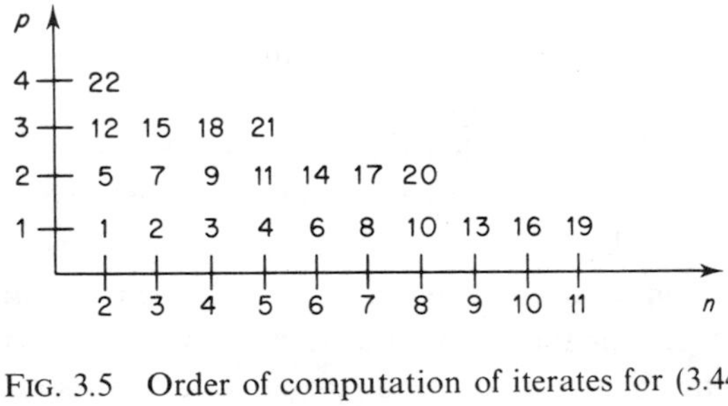

FIG. 3.5 Order of computation of iterates for (3.44).

3.8 Computational aspects of iterative formulae

We now examine the computational aspects of our algorithms in more detail. Following Hull (1974) we find it convenient to regard a numerical method for the solution of a stiff system of equations as consisting of four basic stages; namely

(1) estimate the local truncation error;
(2) make decisions – exiting if appropriate;
(3) prepare for a step – choose stepsize and order;
(4) calculate an approximation to $\mathbf{y}(x_n + h)$.

The procedure which we consider for the estimation of the local truncation error is one based on a one step, two half steps technique coupled with local Richardson extrapolation. This procedure does have the disadvantage of having a relatively high computational cost, when compared for example with Milne's device, since it requires the implicit set of algebraic equations to be solved three times for each integration step. It does however have the advantage of providing a good error estimate in general even when $h\|\partial\mathbf{f}/\partial\mathbf{y}\|$ is very large. Milne's device, although very cheap computationally, has been found to be much less reliable than local Richardson extrapolation, especially in the transient phase, and so it is rejected for our purposes. The decision whether or not to accept a step as satisfactory is based on a desire to keep the size of the local truncation error at each step less than a prescribed maximum, say $\varepsilon(x)$. Since, when using a procedure which requires more than one initial condition for starting and uses as a secondary iteration a modified Newton scheme with a coefficient matrix which is relatively expensive to evaluate and factorise, changes in the steplength of integration are computationally expensive to perform, we shall only make a change in the steplength if there are considerable gains to be made in efficiency. A procedure which we have found to be convenient in practice is to restrict all stepchanges either to doubling or to halving. The steplength is halved if the estimated error at any steppoint x is greater than $\varepsilon(x)$, with the relevant approximation being re-computed using the smaller step, and the steplength is doubled if the estimated error on a doubled step is less than $\frac{1}{2}\varepsilon(x)$. We note that the solution using a doubled step has already been calculated and so it is a trivial problem to estimate the local error in this solution. If at any stage it is found to be necessary to halve the steplength of integration, the procedure adopted in practice with a scheme of stepnumber greater than 1 is to continue the calculation using the variable step form of the particular integration procedure being used. If, for example, it were found to be necessary to change from h to $\frac{1}{2}h$ at the steppoint x_n using scheme (3.40), we would require a formula using the steppoints x_{n-1}, x_n, $x_{n+1/2}$ and x_{n+1}. This variation of scheme (3.40) is very straightforward to derive and will not be given explicitly. Another possibility at points where the step changes would be to continue using a scheme based on equally spaced points, with the spacing of the points being reduced from h to $\frac{1}{2}h$, and to compute the required extra solution values with a sufficiently high order interpolating polynomial. This in turn suggests that it may be advantageous to extend the approach by Nordsieck (1962, 1963) to this class of iterative integration procedures. Research is in progress to determine the relative efficiency of these two approaches.

The problem of choosing the order of our scheme has not really arisen so far since in section 3.7 we have derived only one fifth order and one seventh order scheme. Many other similar schemes of varying order could easily be derived, but this does not seem to be worthwhile until the computational aspects of the algorithms proposed in section 3.7 are better understood and more practical experience has been obtained. When this is the case and the facilities which are needed for the design of efficient schemes are more clear, it should be possible to derive a whole class of computationally efficient, highly stable schemes. It should then be possible to produce an algorithm for choosing between integration procedures of differing order.

The final problem which we consider is that of actually calculating and accepting an approximation at the current steppoint. We have already discussed the method and order of computation in some detail and so the remaining problem is to determine a suitable value for the maximum allowable number, μ, of iterates which we are prepared to compute at each steppoint before we decide that the scheme is not converging sufficiently rapidly and we halve h. As already mentioned, stepchanges are generally rather expensive to perform and so it is necessary to allow μ to be reasonably large. If, however, we allow μ to be too large, overflows may occur in unfavourable circumstances due to rapid divergence of the modified Newton iteration scheme. The determination of a value of μ which is in any sense optimal, if possible at all, will need to be done numerically. This is another computational aspect of our algorithm which requires further attention. In practice we set $\mu = 5$ and this seems quite reasonable.

To conclude this section we present some numerical results which were obtained by applying the iterative integration procedure (3.34) to a particular stiff system of ordinary differential equations. The problem which we shall consider is the following one due to Liniger and Willoughby (1967):

$$\frac{dy}{dx} = 0{\cdot}01 - (0{\cdot}01 + y + z)((y + 1)(y + 1000)), \quad y(0) = 0,$$
$$\frac{dz}{dx} = 0{\cdot}01 - (0{\cdot}01 + y + z)(1 + z^2), \qquad z(0) = 0, \tag{P1}$$

where the range of integration is $0 \leq x \leq 100$. This problem is not too difficult to integrate numerically since it is only of dimension 2, and so could probably be integrated using almost any standard numerical method without requiring a large amount of computation time. However, we consider it here as it is useful for displaying our numerical techniques. The first important conclusion which we draw from using integration schemes of the type described in this section to integrate stiff systems of ordinary differential equations is that there is generally a considerable increase in computational efficiency to be made if an accurate predictor is used to construct the initial approximations $\{\mathbf{y}_n^{(0)}\}$ as they are needed. At the very first step the following implicit "predictor" is used to give the first approximation $\mathbf{y}_1^{(0)}$ to $\mathbf{y}_1$:

$$\mathbf{y}_1 - \mathbf{y}_0 = \frac{h}{2}\{\mathbf{f}_1 + \mathbf{f}_0\} - \frac{h^2}{12}\{\mathbf{g}_1 - \mathbf{g}_0\}, \tag{3.45}$$

where the algebraic equations for $\mathbf{y}_1$ are solved using a modified Newton iteration scheme iterated to convergence with initial approximation $\hat{\mathbf{y}}_1 = \mathbf{y}_0 + h\mathbf{f}_0$. Having obtained our initial approximation from (3.45) iterated to convergence, we estimate $\mathbf{y}_2$ using the fifth order explicit predictor

$$\mathbf{y}_{n+1} - 32\mathbf{y}_n + 31\mathbf{y}_{n-1} = -h\{14\mathbf{f}_{n-1} + 16\mathbf{f}_n\} + h^2\{-2\mathbf{g}_{n-1} + 4\mathbf{g}_n\}, \tag{3.46}$$

with the value $n = 1$. When these two values have been computed we may use scheme (3.34) with $n = 0$ to calculate another approximation to $\mathbf{y}_1$. Using this approximation and the value of $\mathbf{y}_2$ obtained from (3.46) we may use scheme (3.46) with $n = 2$ to calculate an approximation to $\mathbf{y}_3$. Clearly we may continue in this way using scheme (3.34), with the sequence of iterates calculated in the order defined by Fig. 2.1, iterated to convergence with the predicted values being calculated using scheme (3.46). The algebraic equations arising from the use of (3.34) and (3.45) are solved using a modified Newton iteration procedure, of the type considered earlier, iterated to convergence and an estimate of the local truncation error is obtained at every step using a "two half steps–one ordinary step" procedure coupled with local Richardson extrapolation. This error estimate is used in conjunction with a step control procedure which chooses the step so that the estimated local truncation error, T.E., is less than a specified local tolerance ε. By way of illustration suppose for example that we have already obtained the finally accepted approximation $\mathbf{y}_n$ to $\mathbf{y}(x_n)$ and we wish to calculate an approximation to $\mathbf{y}(x_{n+1})$ at the steppoint $x_{n+1} = x_n + h$ using scheme (3.34). Denoting by $\bar{y}_{n+1}$ the solution obtained at x_{n+1} using (3.34) with a steplength h and by $\mathbf{y}_{n+1}$ that obtained at the same point using (3.34) twice with a step length $h/2$, we have

$$\mathbf{y}_{n+1} = \mathbf{y}(x_{n+1}) + 2\mathbf{c}_1(h/2)^6 + 0(h^7),$$

$$\bar{\mathbf{y}}_{n+1} = \mathbf{y}(x_{n+1}) + \mathbf{c}_1 h^6 + 0(h^7),$$

where $\mathbf{c}_1$ is the principal error function associated with (3.34). From these two relations it follows immediately that if we ignore terms of order $0(h^7)$ we have

$$\text{T.E.} \equiv \tfrac{1}{32}\mathbf{c}_1 h^6 = \tfrac{1}{31}(\bar{\mathbf{y}}_{n+1} - \mathbf{y}_{n+1}) \qquad (3.47)$$

which serves as a computable estimate of the local truncation error in $\mathbf{y}_{n+1}$. In order to control this estimate of the local truncation error we use the following procedure:

(1) If $\|\text{T.E.}\| > \varepsilon$ halve h and go back to the step point x_n.
(2) If $\varepsilon/80 < \|\text{T.E.}\| < \varepsilon$ keep h fixed and continue the integration from x_{n+1}.
(3) If $\|\text{T.E.}\| < \varepsilon/80$ double h and continue from x_{n+1}.

As mentioned previously we have found it necessary to monitor carefully the rate of convergence of our primary iteration scheme. If this rate of convergence is not "sufficiently rapid" it is necessary to recompute the Jacobian matrix and to start again from the last point, x_i, at which convergence has been obtained. If the scheme still fails to converge sufficiently rapidly, it is necessary to halve the step and to start again from the steppoint x_i. This procedure may be continued until convergence is once again achieved within the prescribed maximum allowable number of iterates. Practical experience would seem to indicate that a convenient procedure is to set the maximum allowable number of iterates at about 5, since otherwise the scheme may become rather inefficient and in some cases get out of hand by causing overflows. Also it has been found that if we are in a region where it has been found necessary to reduce h due to our iteration scheme not converging sufficiently rapidly, then it is advisable not to try to increase the steplength again for some time. In practice we have taken this "time" to be five steplengths.

In Table 3.5 we present the results obtained for the approximate numerical integration of problem (P1) using scheme (3.34) together with the high order predictors (3.45) and (3.46). It can be seen that the results obtained at the point $x = 100$ are satisfactory, that the integration is stable throughout and that increasingly accurate results are obtained as more strict local error requirements are imposed.

TABLE 3.5
True Solution at $x = 100$ is: $y = -0{\cdot}99164207$, $z = 0{\cdot}98333636$

Tolerance		0·001	0·0001	0·000001
Initial step		0·01	0·01	0·00015625
Number of steps		71	123	512
Largest steplength		2·56	1·28	1·28
Solution obtained	y:	−0·99164099	−0·99164156	−0·99164207
at $x = 100$	z:	0·98333511	0·98333578	0·98333636

3.9 Another class of iterative schemes

In this section we consider an alternative class of iteration schemes based on the algorithms suggested in Section 2.7. Our main aim is to develop integration procedures which are fully A-stable and which converge to the required solution of our discretisation scheme in only one primary iteration. The ideas which we propose may best be explained by considering some particular examples. In the first of these we consider an algorithm based on the explicit mid-point rule

$$\mathbf{y}_{n+1} - \mathbf{y}_{n-1} = 2h\mathbf{f}_n.$$

Following the approach suggested in Section 2.7 we rewrite this scheme in the form

$$\mathbf{y}_{n+1} - 2\mathbf{y}_n + \mathbf{y}_{n-1} + 2\mathbf{y}_n - 2\mathbf{y}_{n-1} = 2h\mathbf{f}_n$$

and use the iteration scheme

$$\mathbf{y}_n^{(t)} - \mathbf{y}_{n-1}^{(t)} = h\mathbf{f}_n^{(t)} - \tfrac{1}{2}\delta^2\mathbf{y}_n^{(t-1)} \tag{3.48}$$

to generate the required solution.

Suppose now that the finally accepted approximation $\mathbf{y}_{n-1}^{(1)}$ to $\mathbf{y}(x_{n-1})$ has been computed and we wish to calculate an approximation to $\mathbf{y}(x_n)$. The first step of our algorithm is to compute the two iterates $\mathbf{y}_n^{(0)}$ and $\mathbf{y}_{n+1}^{(0)}$ using the backward Euler rule

$$\mathbf{y}_j^{(0)} - \mathbf{y}_{j-1}^{(0)} = h\mathbf{f}_j^{(0)}, \qquad j = n, n+1,$$

where we define $\mathbf{y}_{n-1}^{(0)} \equiv \mathbf{y}_{n-1}^{(1)}$. The implicit algebraic equations occurring for

$\mathbf{y}_n^{(0)}$ and $\mathbf{y}_{n+1}^{(0)}$ are solved using a modified Newton iteration scheme iterated to convergence. As our first iterates $\hat{\mathbf{y}}_n$ and $\hat{\mathbf{y}}_{n+1}$ for use with our Newton iteration scheme we take

$$\hat{\mathbf{y}}_j = \mathbf{y}_{j-1}^{(0)} + h\mathbf{f}_{j-1}^{(0)}, \qquad j = n, n+1.$$

Having obtained the quantities $\mathbf{y}_n^{(0)}$ and $\mathbf{y}_{n+1}^{(0)}$ we may compute

$$\delta^2\mathbf{y}_n^{(0)} \equiv \mathbf{y}_{n+1}^{(0)} - 2\mathbf{y}_n^{(0)} + \mathbf{y}_{n-1}^{(0)}$$

and the final iterate $\mathbf{y}_n^{(1)}$ at x_n may be computed using scheme (3.48) with $t = 1$, where again the algebraic equations for $\mathbf{y}_n^{(1)}$ are solved using a modified Newton iteration scheme iterated to convergence with initial approximation $\mathbf{y}_n^{(0)}$. This scheme, which may be shown to be of order two (Cash, 1977b), may now be continued in a step by step fashion over the whole range of integration. In particular, when integrating from x_n to x_{n+1} we re-define $\mathbf{y}_n^{(0)} = \mathbf{y}_n^{(1)}$ and calculate two new approximations $\mathbf{y}_{n+1}^{(0)}$, $\mathbf{y}_{n+2}^{(0)}$ at x_{n+1}, x_{n+2} using the backward Euler rule iterated to convergence. Scheme (3.48) may then be applied with $t = 1$ and n replaced by $n + 1$ to give the iterate $\mathbf{y}_{n+1}^{(1)}$. If we now apply this process to the scalar test equation $y' = \lambda y$, where for the remainder of this section λ will denote a complex constant with negative real part, we obtain an expression of the form

$$y_n^{(1)} = R(q)y_{n-1}^{(1)},$$

where

$$R(q) = (1 - 2q + \tfrac{1}{2}q^2)/(1 - q)^3, \qquad q = h\lambda.$$

It is now straightforward to show that $|R(q)| \leq 1$ for all q where $\mathrm{Re}(q) \leq 0$ showing that our integration scheme is L-stable in the sense of Ehle (1969).

Before examining the computational aspects of this algorithm we find it convenient to consider as our second example the two-step scheme

$$\mathbf{y}_{n+1} + 4\mathbf{y}_n - 5\mathbf{y}_{n-1} = h\{4\mathbf{f}_n + 2\mathbf{f}_{n-1}\}. \tag{3.49}$$

This particular linear multistep method is rather famous since it has been used by Dahlquist to demonstrate instability. In fact, it is not even zero-stable since one of the roots of the polynomial

$$x^2 + 4x - 5 = 0$$

is -5. As a consequence, unsatisfactory results will be obtained if we attempt to solve (3.49) by a direct method. Following the approach adopted in Section

2.7 we could seek to generate the required solution of this scheme using the iteration

$$\mathbf{y}_n^{(t)} - \mathbf{y}_{n-1}^{(t)} = \frac{h}{6}\{4\mathbf{f}_n^{(t)} + 2\mathbf{f}_{n-1}^{(t)}\} - \frac{\delta^2}{6}\mathbf{y}_n^{(t-1)}. \tag{3.50}$$

We define the iterate $\mathbf{y}_{n-1}^{(0)}$ as $\mathbf{y}_{n-1}^{(0)} \equiv \mathbf{y}_{n-1}^{(1)}$, the finally accepted approximation to $\mathbf{y}(x_{n-1})$ at the steppoint x_{n-1}. Scheme (3.50) with $t = 1$ may now be used as a corrector with a separate implicit scheme being used as a "predictor" to calculate $\mathbf{y}_n^{(0)}$ and $\mathbf{y}_{n+1}^{(0)}$. It would seem that a necessary condition for the overall scheme to be A-stable is that the "predictor" should be A-stable. This in turn implies that the "predictor" should be implicit. Although the phrase "implicit predictor" is in some senses a contradiction of terms, since in the current literature the terms "predictor" and "explicit scheme" are synonomous, it does seem that an implicit predictor–corrector scheme is the best way of describing the algorithm just outlined. We shall therefore continue to use this phrase. It may be shown (Cash, 1977b) that if we use the trapezoidal rule as the predictor followed by scheme (3.50) as the corrector, with all the algebraic equations being solved using a modified Newton iteration scheme iterated to convergence, then the whole procedure is A-stable and has order 3. This procedure may be summarised in the following way:

(a) Suppose that we have calculated the finally accepted approximation $\mathbf{y}_{n-1}^{(1)}$ at x_{n-1} and we wish to calculate an approximation $\mathbf{y}_n^{(1)}$ to $\mathbf{y}(x_n)$. Define an iterate $\mathbf{y}_{n-1}^{(0)}$ as $\mathbf{y}_{n-1}^{(0)} \equiv \mathbf{y}_{n-1}^{(1)}$.

(b) Compute iterates $\mathbf{y}_n^{(0)}, \mathbf{y}_{n+1}^{(0)}$ using the implicit predictor

$$\mathbf{y}_j^{(0)} - \mathbf{y}_{j-1}^{(0)} = \frac{h}{2}\{\mathbf{f}_j^{(0)} + \mathbf{f}_{j-1}^{(0)}\}, \qquad j = n, n+1.$$

Here the implicit algebraic equations defining $\mathbf{y}_j^{(0)}$ are solved using a modified Newton method iterated to convergence and the initial iterate used with these iteration schemes is $\hat{\mathbf{y}}_j = \mathbf{y}_{j-1}^{(0)} + h\mathbf{f}_{j-1}^{(0)}, j = n, n+1$.

(c) Compute the final iterate $\mathbf{y}_n^{(1)}$ using scheme (3.50) with $t = 1$. Here the algebraic equations defining $\mathbf{y}_n^{(1)}$ are solved using a modified Newton method iterated to convergence with initial iterate $\mathbf{y}_n^{(0)}$.

Scheme (3.50) is more of theoretical than practical interest since, as we shall see, it in effect breaks both of Dahlquist's famous stability bounds. We note first of all that scheme (3.50) is effectively a one-step method (see Definition 3.1), since it requires only one initial condition of the form $\mathbf{y}(x_0) = \mathbf{y}_0$, and yet it has order 3 whereas the greatest attainable order of a conventional zero-stable one-step method is 2. Scheme (3.50) also breaks Dahlquist's second stability bound since it is A-stable and has order of accuracy greater than two. It cannot be emphasised too strongly at this stage that the algorithms developed in this section bear little relation to general linear multistep methods, but use them merely as a convenient starting point. Since we are not iterating our primary iteration scheme to convergence, but are instead performing a fixed number of primary iterations (where this number has been taken to be 1 in our examples so far), the final sequence of values obtained will not in general satisfy a linear multistep method and so there will be no contradiction with Dahlquist's theorems. We are not attempting to disprove Dahlquist's theorems but are trying to get round the conditions they impose by not iterating our schemes to convergence. As we shall show later it is possible to obtain useful results using more than one primary iteration and in some cases good results can be obtained, for non-stiff problems at least, if we iterate both our primary and secondary iteration schemes to convergence (Miller, 1966).

For the purposes of this section we shall divide the proposed iterative algorithms into two distinct sub-classes; those which may be regarded as genuine implicit predictor–corrector schemes which use distinct predictors and correctors, and those which use the same sort of scheme for both the predictor and the corrector. We consider first of all the second class of schemes which is typified by the algorithm based on (3.48). It is of interest to note that this scheme, and most other schemes belonging to this particular class, has a close connection with the method of deferred (or difference) correction developed by Fox (1957). This connection will be exploited in the forthcoming analysis. In our example the term $-\frac{1}{2}\delta^2\mathbf{y}_n^{(0)}$ serves as a computable approximation to the principal term in the local truncation error of the "$t = 1$" iteration scheme which is, of course, the backward Euler rule. A slight difference is that we apply our primary iteration scheme only once in a precisely defined way, whereas deferred correction is usually applied in a somewhat different way and, in particular, is often iterated until convergence. Nevertheless as we shall see there is still a close connection between these two approaches.

We now consider the computational aspects of scheme (3.48) in more detail and examine the work involved in stepping from x_{n-1} to x_n. We note first of all that in order to solve for $\mathbf{y}_n^{(0)}$ it will be necessary to have available an approximation, $\mathbf{A}_n$, to the matrix $[\mathbf{I} - h(\partial\mathbf{f}(\mathbf{y}_n)/\partial\mathbf{y})]$ for use with our modified Newton iteration scheme. We shall not usually store this matrix as a square array but rather as an $\mathbf{L}_n\mathbf{U}_n$ product of triangular matrices since the systems of linear algebraic equations resulting from our modified Newton iteration scheme can then be solved by simple forward and backward elimination. When computing $\mathbf{y}_{n+1}^{(0)}$ we can often use the same approximation $\mathbf{A}_n$, and hence the same $\mathbf{L}_n\mathbf{U}_n$ factorisation, for $[\mathbf{I} - h(\partial\mathbf{f}(\mathbf{y}_{n+1})/\partial\mathbf{y})]$ and this usually leads to a considerable saving in computational effort. The final stage of our algorithm is to compute the term $\delta^2\mathbf{y}_n^{(0)}$ and then to use (3.48) with $t = 1$ to give $\mathbf{y}_n^{(1)}$. In order to solve for $\mathbf{y}_n^{(1)}$ we again use a modified Newton iteration scheme iterated to convergence. The matrix $\mathbf{A}_n$ serves as our approximation to $[\mathbf{I} - h(\partial\mathbf{f}(\mathbf{y}_n)/\partial\mathbf{y})]$, and as a first approximation to $\mathbf{y}_n^{(1)}$ for use with our modified Newton scheme we use $\mathbf{y}_n^{(0)}$. Since $\mathbf{y}_n^{(0)}$ is usually a reasonably good approximation to $\mathbf{y}_n^{(1)}$, convergence generally occurs fairly rapidly. Having obtained our final approximation $\mathbf{y}_n^{(1)}$ we set $\bar{\mathbf{y}}_{n+1} = \mathbf{y}_{n+1}^{(0)}$ for storage purposes since, as we shall see, this quantity will be used during the next step. In order to calculate our final approximation at the next step we set $n = n + 1$ and carry out exactly the same process. We first of all set $\mathbf{y}_n^{(0)} \equiv \mathbf{y}_n^{(1)}$. We also need to calculate a new approximation $\mathbf{y}_{n+1}^{(0)}$ (i.e. we do not use the approximation $\mathbf{y}_{n+1}^{(0)}$ obtained during the previous step). However, as our first approximation for use with the iteration scheme for $\mathbf{y}_{n+1}^{(0)}$ we use $\mathbf{y}_{n+1}^{(0)} = \bar{\mathbf{y}}_{n+1}$ and this usually ensures that the iteration scheme converges fairly rapidly. When computing the iterates $\mathbf{y}_{n+1}^{(0)}, \mathbf{y}_{n+2}^{(0)}$ and $\mathbf{y}_{n+1}^{(1)}$ we again use the matrix $\mathbf{A}_n$ as our coefficient matrix. If our modified Newton iteration schemes converge to the required degree of precision in five iterations or less we proceed with $\mathbf{A}_n$ still serving as the coefficient matrix of our iteration scheme. If, however, at some point, $m + 1$ say, one of our iteration schemes fails to converge in 5 iterations it will be necessary to compute the matrix $\mathbf{A}_m \equiv [\mathbf{I} - h(\partial\mathbf{f}(\mathbf{y}_m^{(1)})/\partial\mathbf{y})]$ at the last converged value $\mathbf{y}_m^{(1)}$ and to use $\mathbf{A}_m$ instead of $\mathbf{A}_n$. Since this method of using modified Newton iteration schemes is now well known, we shall not consider it any further. From now on we shall refer to this class of schemes which use the same predictor and corrector as Class (B) and we shall return to it at a later stage.

We now examine the first class of methods, which we denote by Class (A), which use separate predictors and correctors and which is typified by (3.50). We see that in order to calculate $\mathbf{y}_n^{(0)}$ using (3.50) it is necessary to compute an approximation to the matrix $[\mathbf{I} - \frac{1}{2}h(\partial\mathbf{f}(\mathbf{y}_n)/\partial\mathbf{y})]$ and to factorise it into an **LU** product. To solve for $\mathbf{y}_n^{(1)}$, however, we need to compute an approximation to $[\mathbf{I} - \frac{2}{3}h(\partial\mathbf{f}(\mathbf{y}_n)/\partial\mathbf{y})]$ and also to obtain the corresponding factorisation. In view of this, more computational effort is in general required with methods of Class (A) than is required with methods of Class (B). In order to compensate for this, however, it may be possible to derive higher order schemes of Class (A) than of

Class (B). In this section we shall be mainly interested in carrying out an analysis of schemes of Class (B) since at the present stage of development this would appear to be the more efficient of the two classes. Schemes of Class (A) do however seem to merit further attention at some future stage.

We also note that it is possible to rewrite our one step iterative methods as diagonally implicit Runge–Kutta formulae. For example scheme (3.48) may be rewritten using Butcher's notation (Lambert, 1973, Chapter 4) as

$$\begin{array}{c|ccc} 1 & 1 & & \\ 2 & 1 & 1 & \\ 1 & \frac{1}{2} & -\frac{1}{2} & 1 \\ \hline & \frac{1}{2} & -\frac{1}{2} & 1 \end{array}$$

From this relation the order of our scheme and the expression for $R(q)$, which contains information regarding the stability properties of the scheme, follow in a straightforward fashion. However our formulae have some important advantages over conventional implicit Runge–Kutta formulae. For a discussion of these the reader is referred to a recent thesis by J. Bond (On a Class of Block Iterative Methods for the Numerical Solution of Stiff Systems of Ordinary Differential Equations, Ph.D. Thesis, University of London, 1979). The formulae, based on our iterative approach, given by Bond seem to be generally superior to conventional Runge–Kutta methods.

We can also rewrite our formulae as non-standard linear multistep methods. If, for example, we consider the iteration scheme (3.50) with the implicit trapezoidal rule used as a predictor, we have

$$\mathbf{y}_n^{(1)} - \mathbf{y}_{n-1}^{(1)} = h\{-\tfrac{1}{12}\mathbf{f}_{n+1}^{(0)} + \tfrac{2}{3}\mathbf{f}_n^{(1)} + \tfrac{5}{12}\mathbf{f}_{n-1}^{(1)}\}. \qquad (3.51)$$

It is of interest to note that iteration scheme (3.51) is precisely iteration scheme (3.12) with $t = 1$. The outcome of our analysis is that if we use scheme (3.12) with just one primary iteration we obtain an A-stable third order scheme providing we construct an appropriate sequence of initial approximations $\mathbf{y}_n^{(0)}$. This sequence is precisely that obtained using two steps of the trapezoidal rule starting from a known value $\mathbf{y}_n^{(1)}$ at x_{n-1} and iterating to convergence.

We mentioned previously that the schemes derived so far in this section are of more theoretical than practical interest and so we now derive some schemes which are of practical use for the numerical integration of stiff systems of equations. The first procedure is based on the fourth order linear multistep method

$$\mathbf{y}_{n+1} + 9\mathbf{y}_n - 9\mathbf{y}_{n-1} - \mathbf{y}_{n-2} = 6h\{\mathbf{f}_n + \mathbf{f}_{n-1}\}. \qquad (3.52)$$

We shall seek to generate the required solution of (3.52) using the primary iteration scheme

$$\mathbf{y}_n^{(t)} - \mathbf{y}_{n-1}^{(t)} = \frac{h}{2}\{\mathbf{f}(\mathbf{y}_n^{(t)}) + \mathbf{f}(\mathbf{y}_{n-1}^{(t)})\} - \frac{1}{12}\delta^2\mathbf{y}_n^{(t-1)} + \frac{1}{12}\delta^2\mathbf{y}_{n-1}^{(t-1)}. \tag{3.53}$$

As usual we shall perform just one primary iteration and so the finally accepted sequence of approximations will not in general satisfy a conventional linear multistep method. Although scheme (3.52) is a convenient starting point for the derivation of (3.53) we could abolish any connection between (3.52) and (3.53) altogether if we regard (3.53) from the outset as a type of deferred correction scheme. This is possible since the $t - 1$ terms appearing in (3.53) serve as an approximation to the principal part, $\frac{-h^3}{12}y_n'''$, of the local truncation error of the part with iteration index t (the trapezoidal rule). In order to make scheme (3.53) $A(\alpha)$-stable for α reasonably large, it is necessary to use quite an elaborate iteration scheme as we shall now explain. We assume first of all that the finally accepted approximations $\mathbf{y}_{n-1}^{(1)}$ and $\mathbf{y}_n^{(1)}$ at the points x_{n-1} and x_n respectively are known and we wish to calculate an approximation to the solution at the steppoint x_{n+1}. To do this we shall use scheme (3.53) in the following step by step fashion.

Step 1
Starting from the finally accepted approximation $\mathbf{y}_{n-1}^{(1)}$ at x_{n-1} calculate approximations $\mathbf{y}_n^{(0)}$, $\mathbf{y}_{n+1}^{(0)}$ and $\mathbf{y}_{n+2}^{(0)}$ using the trapezoidal rule with the associated modified Newton schemes being iterated to convergence. Note that we already have the finally accepted approximation $\mathbf{y}_n^{(1)}$ at x_n but the first part of our algorithm involves the computation of a different approximation $\mathbf{y}_n^{(0)}$ to $\mathbf{y}(x_n)$. We shall comment on this unusual approach at a later stage.

Step 2
Compute the quantity $\delta^2\mathbf{y}_{n+1}^{(0)} - \delta^2\mathbf{y}_n^{(0)} \equiv \mathbf{y}_{n+2}^{(0)} - 3\mathbf{y}_{n+1}^{(0)} + 3\mathbf{y}_n^{(0)} - \mathbf{y}_{n-1}^{(0)}$.

Step 3
Put $t = 1$ in (3.53) and solve the equation

$$\mathbf{y}_{n+1}^{(1)} - \mathbf{y}_n^{(1)} = \frac{h}{2}\{\mathbf{f}_{n+1}^{(1)} + \mathbf{f}_n^{(1)}\} - \frac{1}{12}\{\delta^2\mathbf{y}_{n+1}^{(0)} - \delta^2\mathbf{y}_n^{(0)}\}$$

for the required solution $\mathbf{y}_{n+1}^{(1)}$ using a modified Newton iteration scheme iterated to convergence.

Step 4
Take $\mathbf{y}_{n+1}^{(1)}$ as the finally accepted approximation to $\mathbf{y}_{n+1}$.

Step5
If we have not reached the end of the range of integration compute a sequence $\bar{\mathbf{y}}_{n+j}^{(0)}, j = 1, 2, 3$, using the trapezoidal rule starting from $\mathbf{y}_n^{(1)}$ and using initial approximations $(\mathbf{y}_{n+1}^{(0)}, \mathbf{y}_{n+2}^{(0)}, \mathbf{y}_{n+2}^{(0)})$ to $(\bar{\mathbf{y}}_{n+1}^{(0)}, \bar{\mathbf{y}}_{n+2}^{(0)}, \bar{\mathbf{y}}_{n+3}^{(0)})$ respectively. Set $\mathbf{y}_{n+j}^{(0)} = \bar{\mathbf{y}}_{n+j}^{(0)}, j = 1, 2, 3$, and go to step 2 with n replaced by $n + 1$.

This algorithm may now be applied in a step by step manner for increasing integer values of n to give the required solution over the complete range of integration. It may be shown (Cash, 1977b) that this scheme has order 4 and if we apply it to the scalar test equation $y' = \lambda y$ we obtain the relation

$$y_{n+1}^{(1)} = c_1 y_n^{(1)} + c_0 y_{n-1}^{(1)},$$

where $c_1 = (1 + \frac{1}{2}q)/(1 - \frac{1}{2}q)$, $c_0 = (-q^3)/(12(1 - \frac{1}{2}q)^4)$, $q = h\lambda$. It can be seen from this expression that our scheme yields the approximation given by the trapezoidal rule with an extra term $c_0 y_{n-1}^{(1)}$ added on. The region of absolute stability of our algorithm may now be determined using Schur's theorem (Lambert, 1973, p. 78), from which we may deduce that any point q lies inside

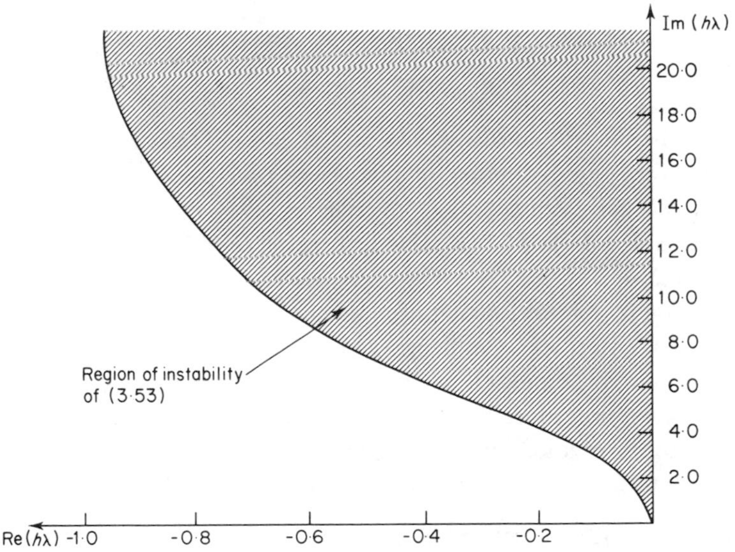

FIG. 3.6 Region of instability of scheme (3.53).

the region of absolute stability of our scheme if, and only if,

(i) $|c_0| < 1$

(ii) $|c_0c_1^* - c_1| < |1 - c_0c_0^*|$

where * denotes the complex conjugate.

The region of the complex left hand half plane where this condition fails to hold (i.e. the region of instability of (3.53)) may be determined using a numerical approach and is shown in Fig. 3.6. As can be seen the algorithm based on (3.53) is $A(\alpha)$-stable for a reasonably large value of the angle α. In order to be able to use this algorithm in practice we need one extra initial condition of the form $\mathbf{y}(x_1) = \mathbf{y}_1$. Since our overall scheme will be of order 4 we shall also require the starting procedure used to generate y_1 to be of order 4. For the purposes of this section we shall use the starting procedure

$$\mathbf{y}_{n+1} - \mathbf{y}_n = h\{\tfrac{1}{2}(\mathbf{k}_1 + \mathbf{k}_4) - \tfrac{1}{12}(\mathbf{k}_2 + \mathbf{k}_5) + \tfrac{1}{12}(\mathbf{k}_3 + \mathbf{k}_6)\}, \tag{3.54}$$

where

$$\begin{aligned}
\mathbf{k}_1 &= \mathbf{f}(\mathbf{y}_{n+1}),\\
\mathbf{k}_2 &= \mathbf{f}(\mathbf{y}_{n+1} + \tfrac{1}{2}h\mathbf{k}_1),\\
\mathbf{k}_3 &= \mathbf{f}(\mathbf{y}_{n+1} - \tfrac{1}{2}h\mathbf{k}_1),\\
\mathbf{k}_4 &= \mathbf{f}(\mathbf{y}_n),\\
\mathbf{k}_5 &= \mathbf{f}(\mathbf{y}_n - \tfrac{1}{2}h\mathbf{k}_4),\\
\mathbf{k}_6 &= \mathbf{f}(\mathbf{y}_n + \tfrac{1}{2}h\mathbf{k}_4),
\end{aligned}$$

since it can be shown that this scheme is A-stable and has order 4. In order to generate the required solution of this scheme we use a modified Newton iteration scheme iterated to convergence. We shall not discuss the computational aspects of this scheme here since they are fully discussed elsewhere (Cash, 1978b).

The precise reason for calculating the sequence of iterates in the described order requires further comment. Assuming that the finally accepted approximations $\mathbf{y}_{n-1}^{(1)}$ and $\mathbf{y}_n^{(1)}$ at the points x_{n-1} and x_n are known, a more straightforward procedure would be to calculate approximations $\mathbf{y}_{n+1}^{(0)}$ and $\mathbf{y}_{n+2}^{(0)}$ using the trapezoidal rule starting from the known solution $\mathbf{y}_n^{(1)}$. Step 2 of our procedure could then be replaced by the computation of

$$\delta^2\mathbf{y}_{n+1}^{(0)} - \delta^2\mathbf{y}_n^{(0)} = \mathbf{y}_{n+2}^{(0)} - 3\mathbf{y}_{n+1}^{(0)} + 3\mathbf{y}_n^{(1)} - \mathbf{y}_{n-1}^{(1)}$$

and step 3 would go through as before. This procedure, which we refer to as

Scheme (D), will in general involve less computational effort than our original scheme (Scheme (C)) but unfortunately there is generally a considerable loss in the degree of accuracy obtained. This is due to the fact that with Scheme (C) the most significant of our error terms combine so as to cancel each other out (see Cash, 1977b) whereas this cancellation does not occur with Scheme (D). It is also possible to use the straightforward deferred correction approach developed by Fox but this would seem to be inappropriate since in general there is a loss in accuracy compared with Scheme (C). Also, if used in the conventional way, it does not allow stepchanges to be made easily. It is clear, however, that there is a close connection between all these approaches, and that Scheme (C) may be regarded as an attempt to extend deferred correction to the numerical solution of stiff systems by performing the iterations in a certain precisely defined order. We shall return to this connection later.

We now illustrate these various approaches by considering an example. As usual we will consider a relatively simple problem since it is easier to evaluate the performances of the various algorithms in this case. For a selection of results obtained for the integration of problems which are more representative of those which arise in practice the reader is referred to Cash (1977b). The problem which we consider is the integration of the linear equation with constant coefficients.

$$\mathbf{y}' = \mathbf{A}\mathbf{y}, \qquad \mathbf{y}(x_0) = \mathbf{y}_0,$$

where

$$\mathbf{y} = \begin{bmatrix} y_1 \\ y_2 \end{bmatrix}, \quad \mathbf{y}_0 = \begin{bmatrix} 1 \\ -1 \end{bmatrix}, \quad x_0 = 0,$$

$$\mathbf{A} = \begin{bmatrix} 0 & 1 \\ -10000 & -10001 \end{bmatrix}.$$

The results obtained for the integration of this problem in various sections of the region of integration are given in Table 3.6. It was found numerically that $y_1 = -y_2$ so we only give the value of y_1 in Table 3.6. It can be seen from this table that both Schemes (C) and (D) give an improvement on the trapezoidal rule and also that both schemes are highly stable. It can also be seen, especially from an examination of the results obtained with $h = 0{\cdot}1$, that Scheme (C) gives a much more substantial increase in accuracy than does Scheme (D). As a result practical experience has shown that it is generally more efficient to use Scheme (C), which is computationally more expensive than Scheme (D) but which obtains greater degrees of accuracy than are achieved by Scheme (D). As

TABLE 3.6

x	h	Local trapezoidal rule	Scheme (C)	Scheme (D)	True solution
0·001	0·001	0·9980019985	0·9980019987	0·9880019987	0·9980019987
0·002	0·001	0·9970044953	0·9970044955	0·9970044955	0·9970044955
0·003	0·001	0·9960079892	0·9960079893	0·9960079893	0·9960079893
1·02	0·02	0·3534542107	0·3534546820	0·3534546618	0·3534546820
1·04	0·02	0·3464553484	0·3464558104	0·3464557723	0·3464558104
1·06	0·02	0·3395950729	0·3395955257	0·3395954706	0·3395955256
1·1	0·1	0·3011439416	0·3011942970	0·3011918623	0·3011942119
1·2	0·1	0·2724863066	0·2725319470	0·2725277470	0·2725317930
1·3	0·1	0·2465558758	0·2465971729	0·2645915477	0·2465969639

before we may rewrite Scheme (C), which was described earlier in step by step form, as a composite non-standard linear multistep method by eliminating $\mathbf{y}_n^{(0)}$, $\mathbf{y}_{n+1}^{(0)}$ and $\mathbf{y}_{n+2}^{(0)}$. The resulting iteration scheme is given by

$$\mathbf{y}_{n+1}^{(1)} - \mathbf{y}_n^{(1)} = \frac{h}{2}\{\mathbf{f}_{n+1}^{(1)} + \mathbf{f}_n^{(1)}\} - \frac{h}{24}\{\mathbf{f}_{n+2}^{(0)} - \mathbf{f}_{n+1}^{(0)} - \mathbf{f}_n^{(0)} + \mathbf{f}_{n-1}^{(0)}\} \quad (3.55a)$$

and it can be seen immediately that this scheme has a local truncation error with asymptotic expansion containing only odd powers of h.

The fairly simple schemes considered so far serve to illustrate some of the basic ideas behind our general approach. These intuitive ideas may, however, be based on a much more sound theoretical basis and for the details the reader is referred to Cash (1978c). Also described in Cash (1978c) is the extension of our approach to iterative schemes based on standard backward differentiation formulae. If, for example, we consider the backward Euler rule

$$\mathbf{y}_{n+1} - \mathbf{y}_n = h\mathbf{f}_{n+1}$$

it can be shown that the asymptotic expansion for the local truncation error of this scheme in powers of the steplength h takes the form

$$-\tfrac{1}{2}h^2\mathbf{y}''(x_{n+1}) + \tfrac{1}{6}h^3\mathbf{y}'''(x_{n+1}) + \cdots \quad (3.55b)$$

Departing from our usual notation we define an iterate $\mathbf{y}_{n,n}^{(0)}$ as $\mathbf{y}_{n,n}^{(0)} \equiv \mathbf{y}_n$ where $\mathbf{y}_n$ is the finally accepted solution at x_n. We now compute two approximate solutions $\mathbf{y}_{n+1,n}^{(0)}, \mathbf{y}_{n+2,n}^{(0)}$ at the grid points x_{n+1}, x_{n+2} using the backward Euler rule iterated to convergence starting from the initial condition $\mathbf{y}(x_n) = \mathbf{y}_{n,n}^{(0)}$. Assuming that $\mathbf{y}_{n,n}^{(0)}$ is exact we have the relation

$$-\tfrac{1}{2}\{\mathbf{y}_{n+2,n}^{(0)} - 2\mathbf{y}_{n+1,n}^{(0)} + \mathbf{y}_{n,n}^{(0)}\} = -\tfrac{1}{2}h^2\mathbf{y}''(x_{n+1}) + 0(h^3).$$

It now follows immediately that the scheme

$$\mathbf{y}_{n+1} - \mathbf{y}_n = h\mathbf{f}_{n+1} - \tfrac{1}{2}\Delta^2\mathbf{y}^{(0)}_{n,n},$$

where

$$\Delta\mathbf{y}^{(0)}_{n,n} \equiv \mathbf{y}^{(0)}_{n+1,n} - \mathbf{y}^{(0)}_{n,n}$$

has order 2. It may also be shown (Cash, 1978c) that this scheme is L-stable. If, as an alternative procedure, we compute values $\mathbf{y}^{(0)}_{n+1,n}$ $\mathbf{y}^{(0)}_{n+2,n}$, $\mathbf{y}^{(0)}_{n+3,n}$ using the trapezoidal rule iterated to convergence starting from the initial condition $\mathbf{y}(x_n) = \mathbf{y}^{(0)}_{n,n}$ we may approximate the first two terms in (3.55b) by the expression

$$-\tfrac{1}{2}\Delta^2\mathbf{y}^{(0)}_{n,n} + \tfrac{1}{6}\Delta^3\mathbf{y}^{(0)}_{n,n}$$

and it follows immediately that the scheme

$$\mathbf{y}_{n+1} - \mathbf{y}_n = h\mathbf{f}_{n+1} - \tfrac{1}{2}\Delta^2\mathbf{y}^{(0)}_{n,n} + \tfrac{1}{6}\Delta^3\mathbf{y}^{(0)}_{n,n}$$

has order 3. It may also be shown (Cash, 1978c) that this scheme is L-stable. This type of approach may be further extended to produce algorithms based on k-step backward differentiation schemes for all $k \le 6$. To see how this may be done we consider the kth order backward differentiation scheme

$$\sum_{j=0}^{k} \alpha_j \mathbf{y}_{n+j} = h\beta_k \mathbf{f}_{n+k}.$$

The leading term in the expression for the local truncation error of this scheme may be written in the form

$$\mathbf{c}_{k+1}h^{k+1}\mathbf{y}^{(k+1)}(x_n) + 0(h^{k+2}).$$

Suppose now that starting from the finally accepted solution $\mathbf{y}^{(0)}_{n,n}$ at x_n we compute a sequence $\mathbf{y}^{(0)}_{n+i,n}$, $1 \le i \le k+1$, using the straightforward trapezoidal rule iterated to convergence. Using the relation

$$\Delta^{k+1}\mathbf{y}^{(0)}_{n,n} = h^{k+1}\mathbf{y}^{(k+1)}(x_n) + 0(h^{k+2})$$

it follows immediately that the scheme

$$\sum_{j=0}^{k} \alpha_j \mathbf{y}_{n+j} = h\beta_k \mathbf{f}(x_{n+k}, \mathbf{y}_{n+k}) + \mathbf{c}_{k+1}\Delta^{k+1}\mathbf{y}^{(0)}_{n,n}$$

will have a local truncation error of order $0(h^{k+2})$. In a similar fashion we may approximate the first two terms appearing in the expression for the local truncation error of our k^{th} order backward differentiation formula to obtain a scheme of order $k + 2$. For details of this approach the reader is again referred to Cash (1978c).

3.10 Cyclic iterative integration procedures

One of our main aims in examining the various classes of iteration schemes described earlier was to derive formulae which had relatively high orders of accuracy, had an infinite region of absolute stability, were self starting and were efficient for the numerical integration of stiff systems of equations. A combination of these characteristics is of course not possible with conventional linear multistep methods. If we wish to use one of the integration procedures described earlier, it will be necessary first of all to compute "predicted" values at the steppoints $(x_{n+1}, x_{n+2}, \ldots, x_{n+j})$, for some $j \geq 2$, using a separate predictor. In some cases, especially in regions where the steplength of integration is changing rapidly, this may result in a large number of predicted values being required. This is clearly an undesirable situation since these predicted values are computed using an implicit scheme which is computationally expensive. This difficulty may be overcome to a large extent if our basic algorithm is allowed to consist of more than one integration scheme with these schemes being applied in a certain precisely defined order. Integration procedures of this general type have been discussed in Cash (1977c) and one of the main findings was that in order to be able to derive a jth order method we need to use a composite integration procedure consisting of j distinct schemes. In order to explain our approach we shall consider the derivation of one particular scheme since other schemes may be derived in a similar manner.

Suppose now that the finally accepted approximation $\mathbf{y}_n^{(1)}$ has been computed at the steppoint x_n and that we wish to compute the solution of our problem at steppoints x_i, $i > n$. As usual we shall re-define the iterate $\mathbf{y}_n^{(0)}$ as $\mathbf{y}_n^{(0)} \equiv \mathbf{y}_n^{(1)}$. The first step in the *block cyclic scheme* which we describe is to compute predicted values $\mathbf{y}_{n+1}^{(0)}, \mathbf{y}_{n+2}^{(0)}, \mathbf{y}_{n+3}^{(0)}, \mathbf{y}_{n+4}^{(0)}$ using the implicit predictor

$$\mathbf{y}_{n+i}^{(0)} - \mathbf{y}_{n+i-1}^{(0)} = \frac{h}{2}\{\mathbf{f}_{n+i}^{(0)} + \mathbf{f}_{n+i-1}^{(0)}\}, \quad i = 1, 2, 3, 4.$$

As usual the required solution of this scheme is generated using a modified Newton iteration procedure iterated to convergence. At the start of the integration, when no past solution values are available, the first iterate for use with the modified Newton scheme is generated using the straightforward Euler rule. However, as soon as three solution values become available quadratic extrapolation through these three values is used to give the required initial iterate. The asymptotic expansion for the local truncation error, T.E., of the trapezoidal rule takes the form

$$\text{T.E.} = -\tfrac{1}{12}h^3\mathbf{y}^{(''')}(x_n) - \tfrac{1}{24}h^4\mathbf{y}^{(\text{IV})}(x_n) + \cdots$$

and the first two terms in this expansion may be approximated by the expression

$$-\tfrac{1}{12}\Delta^3\mathbf{y}_n^{(0)} + \tfrac{1}{12}\Delta^4\mathbf{y}_n^{(0)}$$

with an error of $0(h^5)$. It follows immediately that the scheme

$$\mathbf{y}_{n+1} - \mathbf{y}_n = \frac{h}{2}\{\mathbf{f}_{n+1} + \mathbf{f}_n\} - \frac{1}{12}\Delta^3\mathbf{y}_n^{(0)} + \frac{1}{12}\Delta^4\mathbf{y}_n^{(0)} \tag{3.56}$$

is of order 4 and it is this scheme which is used to compute the approximate solution at the steppoint x_{n+1}. The required solution $\mathbf{y}_{n+1}^{(1)}$ is obtained using a quasi-Newton iteration scheme iterated to convergence. It may be shown that scheme (3.56), which clearly belongs to Class B described earlier, is A-stable. An estimate of the local truncation error of this scheme may now be derived in a straightforward manner. The basic approach is to derive a locally less accurate solution $\bar{\mathbf{y}}_{n+1}$ using a separate implicit scheme. The difference $\mathbf{y}_{n+1}^{(1)} - \bar{\mathbf{y}}_{n+1}$ then serves as an estimate of the local truncation error in the value $\bar{\mathbf{y}}_{n+1}$. If this is less than a prescribed local tolerance, ε, then local extrapolation is performed and the solution $\mathbf{y}_{n+1}^{(1)}$ is carried forward; that is, the estimate is actually for the error in $\bar{\mathbf{y}}_{n+1}$ but it is the asymptotically more accurate solution $\mathbf{y}_{n+1}^{(1)}$ which is carried forward. If this estimate is greater than ε the steplength is halved and the same procedure is carried out again starting from x_n. The scheme used to compute $\bar{\mathbf{y}}_{n+1}$ is

$$\mathbf{y}_{n+1} - \mathbf{y}_n = \frac{h}{2}\{\mathbf{f}_{n+1} + \mathbf{f}_n\} - \frac{1}{12}\Delta^3\mathbf{y}_n^{(0)} \tag{3.57}$$

and it may be shown that this scheme is of order 3. The required solution of (3.57) is computed using a modified Newton iteration procedure iterated to convergence with initial iterate $\frac{1}{2}\{\mathbf{y}_{n+1}^{(0)} + \mathbf{y}_{n+1}^{(1)}\}$. Since we do not normally need to re-compute the coefficient matrix of this modified Newton scheme, and since our initial iterate is usually a very good approximation to the required solution, the value $\bar{\mathbf{y}}_{n+1}$ may be obtained with relatively little computational effort with our iteration scheme often converging in just one or two iterations. In order to calculate our approximation at the steppoint x_{n+2}

we use another predictor–corrector scheme where we impose the two important restrictions that

(a) we do not want to compute any more predicted values $\mathbf{y}_i^{(0)}$; and

(b) we want to allow ourselves the possibility of not having to factorise any more coefficient matrices.

With these restrictions in mind the scheme which we use to calculate $\mathbf{y}_{n+2}$ is

$$\mathbf{y}_{n+2} - \mathbf{y}_{n+1} = \frac{h}{2}\{\mathbf{f}_{n+2} + \mathbf{f}_{n+1}\} - \frac{1}{12}\Delta^3\mathbf{y}_n^{(0)}. \tag{3.58}$$

We note that this scheme uses the predicted values $\mathbf{y}_n^{(0)}, \ldots, \mathbf{y}_{n+3}^{(0)}$ which have already been computed and so restriction (a) is satisfied. Also the coefficient matrix of the modified Newton iteration scheme used to solve for $\mathbf{y}_{n+2}$ is

$$\mathbf{I} - \frac{h}{2}\left[\frac{\partial \mathbf{f}}{\partial \mathbf{y}}(\mathbf{y}_{n+2})\right],$$

which is of a similar form to the coefficient matrix of the modified Newton scheme used to solve for $\mathbf{y}_{n+1}$, and so condition (b) is satisfied. In order to compute a solution $\bar{\mathbf{y}}_{n+2}$ for use in estimating the local truncation error in the way previously described, we use the third order scheme

$$\mathbf{y}_{n+2} - \mathbf{y}_{n+1} = \frac{h}{2}\{\mathbf{f}_{n+2} + \mathbf{f}_{n+1}\} - \frac{1}{12}\Delta^3\mathbf{y}_{n+1}^{(0)}. \tag{3.59}$$

As our initial iterate for use with the modified Newton iteration scheme we use $\frac{1}{2}\{\mathbf{y}_{n+2}^{(0)} + \mathbf{y}_{n+2}\}$ and exactly the same remarks as were made earlier regarding the relative cheapness of obtaining $\bar{\mathbf{y}}_{n+1}$ now apply to $\bar{\mathbf{y}}_{n+2}$. Following this line of attack and bearing in mind restrictions (a) and (b) we may compute the approximation $\mathbf{y}_{n+3}$ using the scheme

$$\mathbf{y}_{n+3} - \mathbf{y}_{n+2} = \frac{h}{2}\{\mathbf{f}_{n+3} + \mathbf{f}_{n+2}\} - \frac{1}{12}\Delta^3\mathbf{y}_n^{(0)} - \frac{1}{12}\Delta^4\mathbf{y}_n^{(0)}. \tag{3.60}$$

We note once again that the initial approximations $\mathbf{y}_n^{(0)}, \ldots, \mathbf{y}_{n+4}^{(0)}$ have already been computed and the coefficient matrix of our modified Newton iteration scheme is $\mathbf{I} - \frac{h}{2}\left[\frac{\partial \mathbf{f}}{\partial \mathbf{y}}(\mathbf{y}_{n+3})\right]$. This is of a similar form to the coefficient matrices of the modified Newton schemes used to compute $\mathbf{y}_{n+1}$ and $\mathbf{y}_{n+2}$. Thus restrictions (a) and (b) imposed earlier are again satisfied. In order to compute the less accurate solution $\bar{\mathbf{y}}_{n+3}$ for use in our error estimation procedure we use the scheme

$$\mathbf{y}_{n+3} - \mathbf{y}_{n+2} = \frac{h}{2}\{\mathbf{f}_{n+3} + \mathbf{f}_{n+2}\} - \frac{1}{12}\Delta^3\mathbf{y}_n^{(0)}. \tag{3.61}$$

We now consider two possible schemes for the calculation of $\mathbf{y}_{n+4}$. Once this value has been computed we shall regard our block cyclic scheme as being over and we shall start again from x_{n+4}. If we demand that our overall scheme should be $L(\alpha)$-stable for α close to $\pi/2$ it would seem to be necessary for us to use an $L(\alpha)$-stable scheme to compute $\mathbf{y}_{n+4}$. One obvious candidate is the fourth order backward differentiation scheme

$$\mathbf{y}_{n+4} - \left(\frac{48}{25}\right)\mathbf{y}_{n+3} + \left(\frac{36}{25}\right)\mathbf{y}_{n+2} - \left(\frac{16}{25}\right)\mathbf{y}_{n+1} + \left(\frac{3}{25}\right)\mathbf{y}_n = \left(\frac{12h}{25}\right)\mathbf{y}'_{n+4}. \quad (3.62)$$

We denote the composite scheme obtained by using (3.56), (3.58), (3.60) and (3.62) in a block cyclic fashion by scheme (a). The region of stability of scheme (a) may be found in the usual way and is given in Cash (1977c). In order to compute the less accurate solution $\bar{\mathbf{y}}_{n+4}$ suitable for use with our error estimation procedure, we use the scheme

$$\mathbf{y}_{n+4} - \mathbf{y}_{n+3} = \frac{h}{2}\{\mathbf{f}_{n+4} + \mathbf{f}_{n+3}\} - \frac{1}{12}\Delta^3\mathbf{y}^{(0)}_{n+1}. \quad (3.63)$$

As an alterative procedure we could compute $\mathbf{y}_{n+4}$ using a scheme which is $A(\alpha)$-stable (but not $L(\alpha)$-stable) for α close to $\pi/2$. The scheme which we use is

$$\mathbf{y}_{n+4} - \mathbf{y}_{n+3} = \frac{h}{2}\{\mathbf{f}_{n+4} + \mathbf{f}_{n+3}\} - \frac{1}{12}\Delta^3\mathbf{y}^{(0)}_{n} - \frac{1}{6}\Delta^4\mathbf{y}^{(0)}_{n}. \quad (3.64)$$

This again uses the predicted values $\mathbf{y}^{(0)}_n, \ldots, \mathbf{y}^{(0)}_{n+4}$ which have already been computed and yields a block cyclic iteration scheme (which we denote by scheme (b)) which is $A(\alpha)$-stable and which provides us with the possibility of having to factorise at most only one coefficient matrix for each block step forward. The region of absolute stability of our composite scheme may be found numerically in the usual way and it may be shown that scheme (b) is "almost" A-stable with a small region of instability close to the imaginary axis. In order to compute a solution $\bar{\mathbf{y}}_{n+4}$ for use with our error estimation procedure we again use scheme (3.63). We note that our block iteration scheme, as so far described, is designed so that an estimate of the local truncation error is available at each steppoint. If, however, we are integrating a system where the solution is required only at the end point of the range of integration, it may be more efficient when using a block cyclic scheme of the type just described if we attempt only to estimate the error in the solution obtained at the end of each block step; that is, if we use a block cyclic scheme consisting of j distinct formulae (and we shall refer to this as a *block j-cyclic scheme*) we estimate the local truncation error only at every j^{th} step and we do not worry about the size of the error committed at intermediate steps. This second approach, which is not generally as robust as the first approach, but which may often be the more efficient of the two, is described more fully in Cash (1977c).

3.11 Miscellaneous iteration schemes

In this section we mention a selection of miscellaneous techniques which are possible extensions of the techniques derived earlier but which have not been fully investigated so far. The algorithms which we have considered have been specially designed so that

(a) the integration procedure is $A(\alpha)$-stable for α close to $\pi/2$;
(b) the iteration scheme converges in just one primary iteration.

It is also possible, however, to develop schemes which still converge to the required solution in a fixed number of primary iterations but where now this number is greater than 1. The procedure of iterating the primary iteration scheme to convergence would not seem to be consistent with the efficient numerical integration of stiff systems, unless substantial increases in the accuracy of the solution obtained are achieved for each iteration. This is due to the large number of function evaluations that would generally be required if this approach of performing more than one primary iteration were to be adopted. If, however, we allow two primary iterations to be performed, we are able to derive high order schemes with an infinite region of absolute stability which are still efficient for the numerical integration of stiff systems. As an example we consider the derivation of an algorithm based on the backward Euler rule. Suppose now that we have computed the finally accepted approximation $\mathbf{y}_n$ at the steppoint x_n and we wish to compute the solution of our system at the steppoint $x_{n+1} = x_n + h$. The first step of our algorithm is to compute a sequence $\{\mathbf{y}_{n+i}^{(0)}\}$ using the scheme

$$\mathbf{y}_{n+i} - \mathbf{y}_{n+i-1} = h\mathbf{f}_{n+i}, \qquad i = 1, 2, 3, 4, \tag{3.65}$$

where the modified Newton scheme defining the $\mathbf{y}_{n+i}$ is iterated to convergence. The next step is to compute a sequence $\{\mathbf{y}_{n+i}^{(1)}\}$ using the scheme

$$\mathbf{y}_{n+i} - \mathbf{y}_{n+i-1} = h\mathbf{f}_{n+i} - \tfrac{1}{2}\{\mathbf{y}_{n+i+1}^{(0)} - 2\mathbf{y}_{n+i}^{(0)} + \mathbf{y}_{n+i-1}^{(0)}\}, \quad i = 1, 2, 3, \tag{3.66}$$

where again the modified Newton schemes are iterated to convergence. Finally we compute an iterate $\mathbf{y}_{n+1}^{(2)}$ using

$$\mathbf{y}_{n+1}^{(2)} - \mathbf{y}_n^{(2)} = h\mathbf{f}_{n+1}^{(2)} - \tfrac{1}{2}\Delta_h^2\mathbf{y}_n^{(1)} + \tfrac{1}{6}\Delta_h^3\mathbf{y}_n^{(1)}, \tag{3.67}$$

where $\Delta_h\mathbf{y}_n \equiv \mathbf{y}_{n+1} - \mathbf{y}_n$. It may be shown† that this scheme has order 3 and is $A(\alpha)$-stable with α close to $\pi/2$. By arranging this scheme to be the first step of a cyclic scheme we may derive a *block 3-cyclic scheme* which is $A(\alpha)$-stable and which requires about three iterates to be computed at each step. Details of this scheme and numerous refinements and additional high order formulae can be found elsewhere.†

A second modification of our approach is the possibility of deriving an implicit predictor–corrector scheme where the predictor is used on a grid which is finer than that used for the corrector. As an example we consider the corrector

$$\mathbf{y}_{n+1} - \mathbf{y}_n = \frac{h}{2}\{\mathbf{f}_{n+1} + \mathbf{f}_n\}.$$

The local truncation error of this scheme may be written in the form

$$-\frac{h^3}{12}\mathbf{y}'''(x_{n+1/2}) + 0(h^5),$$

and following the approach which was developed in previous sections we shall seek to approximate this term using a suitable finite difference expression.

If we now use the predictor

$$\mathbf{y}_{n+i/3}^{(0)} - \mathbf{y}_{n+(i-1)/3}^{(0)} = \frac{h}{6}\{\mathbf{f}(x_{n+i/3}, \mathbf{y}_{n+i/3}^{(0)}) + \mathbf{f}(x_{n+(i-1)/3}, \mathbf{y}_{n+(i-1)/3}^{(0)})\}, \qquad i = 1, 2, 3,$$

to compute iterates $\mathbf{y}_{n+1/3}^{(0)}, \mathbf{y}_{n+2/3}^{(0)}, \mathbf{y}_{n+1}^{(0)}$ we may approximate the leading term in the local truncation error of the trapezoidal rule by

$$-\tfrac{27}{12}\Delta_{h/3}^3\mathbf{y}_n^{(0)} \equiv -\tfrac{27}{12}\{\mathbf{y}_{n+1}^{(0)} - 3\mathbf{y}_{n+2/3}^{(0)} + 3\mathbf{y}_{n+1/3}^{(0)} - \mathbf{y}_n^{(0)}\}.$$

It now follows immediately that the scheme

$$\mathbf{y}_{n+1} - \mathbf{y}_n = \frac{h}{2}\{\mathbf{y}'_{n+1} + \mathbf{y}'_n\} - \frac{27}{12}\Delta_{h/3}^3\mathbf{y}_n^{(0)} \tag{3.68}$$

† J. Bond, "On a Class of Block Iterative Methods for the Numerical Solution of Stiff Systems of Ordinary Differential Equations". Ph.D. Thesis, University of London, 1979).

has order 4. Applying (3.68) to the scalar test equation $y' = \lambda y$ we obtain an expression of the form $y_{n+1} = R(q)y_n$ where

$$R(q) = \frac{1 - \dfrac{q^2}{6} - \left(\dfrac{5}{108}\right)q^3 - \left(\dfrac{1}{432}\right)q^4}{1 - q + \dfrac{q^2}{3} - \left(\dfrac{5}{108}\right)q^3 + \left(\dfrac{1}{432}\right)q^4}.$$

It may be verified that $|R(q)| \leq 1$ for all q where $\mathrm{Re}(q) \leq 0$ showing that this scheme is A-stable. In order to obtain an estimate of the local truncation error of this scheme we compute a solution $\hat{\mathbf{y}}_{n+1}$ using a third order scheme of the form

$$\mathbf{y}_{n+1} - \mathbf{y}_n = \frac{h}{2}\{\mathbf{f}_{n+1} + \mathbf{f}_n\} + h \sum_{i=0}^{3} \beta_i \mathbf{f}^{(0)}_{n+i/3}, \tag{3.69}$$

where $\mathbf{f}_n^{(0)} \equiv \mathbf{f}(x_n, \mathbf{y}_n^{(0)})$. This can in general be done in an infinite number of ways since the condition that (3.69) should be of exact order 3 gives three linear equations in four unknowns. The difference $\varepsilon_{n+1} = \mathbf{y}_{n+1} - \hat{\mathbf{y}}_{n+1}$ now gives an estimate of the local truncation error in $\hat{\mathbf{y}}_{n+1}$. Assuming that there is a local error tolerance T.E. imposed at each step, and letting h_n and h_n' denote the current steplength and the next steplength to be used respectively, the steplength of integration is controlled in the following way:

(a) If $|\varepsilon_{n+1}| > \text{T.E.}$ $h_n' = \frac{1}{2}h_n$ and start again from x_n.
(b) If $\text{T.E.}/20 < |\varepsilon_{n+1}| < \text{T.E.}$, $h_n' = h_n$.
(c) If $|\varepsilon_{n+1}| < \text{T.E.}/20^i$, $h_n' = 2^i h_n$, $i = 1, 2, 3, 4$,

where $|\varepsilon_{n+1}| \equiv \|\varepsilon_{n+1}\|_\infty$. Here the factor 20 is introduced to provide us with some margin of error so that the steplength is not doubled too often when it is not safe for this to be done. If the current step from x_n to x_{n+1} is acceptable, i.e. $|\varepsilon_{n+1}| < \text{T.E.}$, then local extrapolation is performed and it is the solution $\mathbf{y}_{n+1}$ which is actually carried forward. The order relations arising from the condition that (3.69) should be of order 3 are

$$0 = \beta_3 + \beta_2 + \beta_1 + \beta_0,$$

$$0 = \beta_3 + (\tfrac{2}{3})\beta_2 + (\tfrac{1}{3})\beta_1,$$

$$-\tfrac{1}{12} = \beta_3/2 + (\tfrac{2}{9})\beta_2 + (\tfrac{1}{18})\beta_1.$$

There are clearly infinitely many solutions to this system with one particular solution being

$$\beta_3 = -\tfrac{1}{4}, \beta_2 = 0, \beta_1 = \tfrac{3}{4}, \beta_0 = -\tfrac{1}{2}.$$

This particular choice of coefficients produces the scheme

$$\mathbf{y}_{n+1} - \mathbf{y}_n = \frac{h}{2}\{\mathbf{y}'_{n+1} + \mathbf{y}'_n\} - \frac{3h}{4}\{\tfrac{1}{3}\mathbf{f}^{(0)}_{n+1} - \mathbf{f}^{(0)}_{n+1/3} + \tfrac{2}{3}\mathbf{f}^{(0)}_n\}$$

and it is this scheme which produces $\hat{\mathbf{y}}_{n+1}$ for use in the estimation of the local truncation error of scheme (3.68).

A final variation of our approach which we shall now consider is the possibility of allowing our correction term to consist of both solution values and derivative values. To introduce this new approach and to illustrate why, in certain cases, it has advantages compared with the approach already considered, we shall confine our attention to a particular scheme which is suggestive of the general approach. We consider again scheme (3.49) which is given by

$$\mathbf{y}_{n+1} + 4\mathbf{y}_n - 5\mathbf{y}_{n-1} = h\{4\mathbf{f}_n + 2\mathbf{f}_{n-1}\}.$$

In order to generate the required solution of this scheme we previously used the iteration

$$\mathbf{y}^{(1)}_n - \mathbf{y}^{(1)}_{n-1} = \frac{h}{6}\{4\mathbf{f}^{(1)}_n + 2\mathbf{f}^{(1)}_{n-1}\} - \frac{\delta^2}{6}\mathbf{y}^{(0)}_n,$$

where the $\mathbf{y}^{(0)}_n$ terms were computed using the trapezoidal rule

$$\mathbf{y}^{(0)}_{n+1} - \mathbf{y}^{(0)}_n = \frac{h}{2}\{\mathbf{f}^{(0)}_n + \mathbf{f}^{(0)}_{n+1}\}.$$

If we consider generating the required solutions of these two schemes using a modified Newton iteration procedure it will be necessary with the first scheme to factorise the coefficient matrix $\mathbf{I} - \frac{2}{3}h\mathbf{J}_n$ into an **LU** decomposition and with the second scheme we will need to factorise $\mathbf{I} - \frac{1}{2}h\mathbf{J}_n$ into a different **LU** decomposition, where $\mathbf{J}_n$ is a piecewise constant approximation to the relevant Jacobian matrix. Clearly there would in general be a considerable saving in computational effort if we could arrange for the coefficient matrices to be the same in both cases since then only one **LU** decomposition would be required. Suppose now that instead of using iteration (3.50) we use

$$\mathbf{y}^{(1)}_n - \mathbf{y}^{(1)}_{n-1} = \frac{h}{2}\{\mathbf{f}^{(1)}_n + \mathbf{f}^{(1)}_{n-1}\} + \frac{h}{6}\{\mathbf{f}^{(0)}_n - \mathbf{f}^{(0)}_{n-1}\} - \frac{\delta^2}{6}\mathbf{y}^{(0)}_n.$$

We then achieve our aim of having just one iteration matrix to decompose into an **LU** product. It is straightforward to verify that if the trapezoidal rule is used to generate the terms of iteration index 0 our scheme has order 3 and when applied to the scalar test equation $y' = \lambda y$, λ being a complex constant with negative real part, it yields the relation

$$\frac{y_{n+1}}{y_n} = \frac{1 - \dfrac{q}{2} - \dfrac{q^2}{4} + \dfrac{1}{24}q^3}{1 - \dfrac{3q}{2} + \dfrac{3q^2}{4} - \dfrac{q^3}{8}},$$

where $q = h\lambda$. It is trivial to verify that $\left|\dfrac{y_{n+1}}{y_n}\right| < 1$ for all q with $\mathrm{Re}(q) < 0$, thus showing this scheme to be A-stable. As yet no systematic study of this new approach has been carried out, and this is yet another aspect of our iterative method of approach which seems to merit further research.

3.12 Second order equations

In this section we mention the possible extension to the numerical solution of second order ordinary differential equations of some of the techniques developed in previous sections with a view to using these extended techniques to solve two-point boundary value problems. We will be concerned mainly with the linear case where, in order to simplify our analysis, we will assume that the first derivative of y is absent. This is a reasonable assumption to make in this case since we may remove the first derivative of y, if it is present initially, by means of a simple transformation. The problem which concerns us is the solution of the linear equation

$$y'' = f(x)y, \tag{3.70}$$

where the boundary conditions take one of the two forms

$$y(x_0) = y_0, \qquad y(x_N) = y_N, \tag{3.71a}$$

or

$$y(x_0) = y_0, \qquad y(x) \to 0 \text{ as } x \to \infty. \tag{3.71b}$$

It is of course possible for other conditions apart from (3.71b) to be imposed at "infinity", but this is one of the most common boundary conditions of this type. Also, since the proposed technique can usually be extended to deal with most other kinds of linear conditions imposed at infinity, we shall confine our attention to (3.71b) when discussing this class of problem. We will consider two particular methods for the numerical solution of (3.70), one being basically an initial value approach for use when the conditions take the form (3.71b) and one being a genuine boundary value approach for use when the conditions are of the form (3.71a).

We consider first an approach which is applicable when the boundary conditions take the form typified by (3.71b). Any solution $y(x)$ of (3.70) may be written in the form

$$y(x) = c_1 y_1(x) + c_2 y_2(x),$$

where $y_1(x)$, $y_2(x)$ are any two solutions of (3.70) which we assume can be chosen so that any boundary conditions imposed on (3.70) can be satisfied and c_1, c_2 are arbitrary constants. We will assume firstly that it is possible to choose these basis solutions such that

$$\lim_{x \to \infty} \frac{y_2(x)}{y_1(x)} = 0$$

and secondly that we require a solution of (3.70) which is directly proportional to $y_2(x)$. We further assume that $|y_2(x)|$ is monotonically decreasing. The procedure which we adopt is to replace the second derivative appearing in (3.70) by a suitable finite difference approximation and to solve the resulting linear recurrence relation by one of the indirect methods considered in Chapter 2. Suppose now that we approximate the second derivative using an expression of the form

$$y''_{n+1} = \frac{1}{h^2} \sum_{i=0}^{k} \alpha_i y_{n+i}. \tag{3.72}$$

Substitution of this into (3.70) produces the relation

$$\sum_{i=0}^{k} \alpha_i y_{n+i} = h^2 f(x_{n+1}) y_{n+1},$$

which we may rewrite in the form

$$\sum_{i=0}^{k} \gamma_i(x_n) y_{n+i} = 0. \tag{3.73}$$

The reason why we make our finite difference approximation to y''_{n+1} rather than to y''_n is simply that, by making an approximation of the form (3.72), it has been found easier to derive schemes which are of practical use. By adopting this approach our problem may be reduced to one of generating the required solution of a linear recurrence relation subject to boundary conditions of the form (3.71b). Since one of our main problems will be to estimate a practical approximation, N say, to infinity we wish to use a procedure which estimates this quantity automatically as the integration is performed. As mentioned previously in Chapter 2 schemes which do require the value of N to be guessed in advance are of rather unsure practical value, since too small a value of N will result in insufficient precision being obtained and too large a value of N will result in an excessive amount of computation being required. A technique for solving linear recurrence relations with boundary conditions of this type has recently been developed by the author (Cash, 1978d). In order to explain how the technique works in this particular context we again find it convenient to consider a particular example. We make the third order approximation

$$y''_{n+1} = -\frac{1}{12h^2}\{y_{n+4} - 4y_{n+3} - 6y_{n+2} + 20y_{n+1} - 11y_n\},$$

to the second derivative. Using this in (3.70) we obtain the relation

$$y_{n+4} - 4y_{n+3} - 6y_{n+2} + [20 + 12h^2 f(x_{n+1})]y_{n+1} - 11y_n = 0. \quad (3.74)$$

The first problem we need to consider is that of convergence; i.e. we need to determine the conditions under which the solution generated by our algorithm is the required solution of (3.74). In order to do this we first need to prove the following lemma.

Lemma 3.2

Consider the numerical solution of the constant coefficient equation

$$\sum_{j=0}^{4} a_j y_{n+j} = 0$$

using the algorithm described in Section 2.5 with $q = 1$. *Then if the roots* ω_i, $i = 1, 2, 3, 4,$ *of the polynomial* $\Sigma_{j=0}^{4} a_j x^j = 0$ *are such that* $|\omega_1| \geq |\omega_2| \geq |\omega_3| > |\omega_4|$, *where* $\omega_1 \neq \omega_2$, $\omega_2 \neq \omega_3$, $\omega_3 \neq \omega_1$, *it follows that the solution* $y_n^{[N]}$ *generated by our algorithm satisfies*

$$\lim_{N\to\infty} y_n^{[N]} = k_1 \omega_4^{n-1} \quad \text{for all } n.$$

Proof

We first note that the general solution of our linear recurrence relation with constant coefficients is

$$y_n = \sum_{i=1}^{4} c_i \omega_i^n.$$

Using the boundary conditions $y_1^{[N]} = k_1$, $y_N^{[N]} = y_{N+1}^{[N]} = y_{N+2}^{[N]} = 0$ we have

$$k_1 = c_1^{[N]}\omega_1 + c_2^{[N]}\omega_2 + c_3^{[N]}\omega_3 + c_4^{[N]}\omega_4,$$

$$0 = c_1^{[N]}\omega_1^N + c_2^{[N]}\omega_2^N + c_3^{[N]}\omega_3^N + c_4^{[N]}\omega_4^N,$$

$$0 = c_1^{[N]}\omega_1^{N+1} + c_2^{[N]}\omega_2^{N+1} + c_3^{[N]}\omega_3^{N+1} + c_4^{[N]}\omega_4^{N+1},$$

$$0 = c_1^{[N]}\omega_1^{N+2} + c_2^{[N]}\omega_2^{N+2} + c_3^{[N]}\omega_3^{N+2} + c_4^{[N]}\omega_4^{N+2}.$$

Solving these equations by Cramer's rule we have

$$c_1^{[N]} = \frac{\begin{bmatrix} k & \omega_2 & \omega_3 & \omega_4 \\ 0 & \omega_2^N & \omega_3^N & \omega_4^N \\ 0 & \omega_2^{N+1} & \omega_3^{N+1} & \omega_4^{N+1} \\ 0 & \omega_2^{N+2} & \omega_3^{N+2} & \omega_4^{N+2} \end{bmatrix}}{D},$$

$$c_2^{[N]} = \frac{\begin{bmatrix} \omega_1 & k & \omega_3 & \omega_4 \\ \omega_1^N & 0 & \omega_3^N & \omega_4^N \\ \omega_1^{N+1} & 0 & \omega_3^{N+1} & \omega_4^{N+1} \\ \omega_1^{N+2} & 0 & \omega_3^{N+2} & \omega_4^{N+2} \end{bmatrix}}{D},$$

$$c_3^{[N]} = \frac{\begin{bmatrix} \omega_1 & \omega_2 & k & \omega_4 \\ \omega_1^N & \omega_2^N & 0 & \omega_4^N \\ \omega_1^{N+1} & \omega_2^{N+1} & 0 & \omega_4^{N+1} \\ \omega_1^{N+2} & \omega_2^{N+2} & 0 & \omega_4^{N+2} \end{bmatrix}}{D},$$

$$c_4^{[N]} = \frac{\begin{bmatrix} \omega_1 & \omega_2 & \omega_3 & k \\ \omega_1^N & \omega_2^N & \omega_3^N & 0 \\ \omega_1^{N+1} & \omega_2^{N+1} & \omega_3^{N+1} & 0 \\ \omega_1^{N+2} & \omega_2^{N+2} & \omega_3^{N+2} & 0 \end{bmatrix}}{D},$$

where

$$D = \begin{bmatrix} \omega_1 & \omega_2 & \omega_3 & \omega_4 \\ \omega_1^N & \omega_2^N & \omega_3^N & \omega_4^N \\ \omega_1^{N+1} & \omega_2^{N+1} & \omega_3^{N+1} & \omega_4^{N+1} \\ \omega_1^{N+2} & \omega_2^{N+2} & \omega_3^{N+2} & \omega_4^{N+2} \end{bmatrix}.$$

Dividing both the numerator and the denominator of the expressions for $c_i^{[N]}$ by $\omega_1^n\omega_2^n\omega_3^n$ we have

$$\lim_{N\to\infty} c_1^{[N]} = \lim_{N\to\infty} c_2^{[N]} = \lim_{N\to\infty} c_3^{[N]} = 0,$$

$$\lim_{N\to\infty} c_4^{[N]} = \frac{(\omega_1 - \omega_2)(\omega_1 - \omega_3)(\omega_2 - \omega_3)k_1}{(\omega_1 - \omega_2)(\omega_1 - \omega_3)(\omega_2 - \omega_3)\omega_4} \neq 0.$$

From these expressions for $c_i^{[N]}$ it follows immediately that

$$\lim_{N\to\infty} y_i^{[N]} = k_1\omega_4^{i-1}.$$

This completes the proof of Lemma 3.2.

It now follows that, providing the roots of our polynomial

$$\sum_{j=0}^{4} a_j x^j = 0$$

satisfy the conditions stated in the lemma, the algorithm described in Section 2.5 (which from now on we shall refer to as algorithm (2.5)) will converge to the minimal solution of the original recurrence relation. The concept of stability of our algorithm may now be developed in the following way. Suppose first of all that we apply scheme (3.74) to the scalar test equation $y'' = \lambda^2 y$, where λ is a real constant, to produce

$$y_{n+4} - 4y_{n+3} - 6y_{n+2} + [20 + 12h^2\lambda^2]y_{n+1} - 11y_n = 0.$$

The general solution of this recurrence relation takes the form

$$y_n = \sum_{i=1}^{4} c_i[\omega_i(\lambda)]^n,$$

where the $\omega_i(\lambda)$ are the four roots, assumed distinct, of the polynomial

$$x^4 - 4x^3 - 6x^2 + [20 + 12q]x - 11 = 0, \tag{3.75a}$$

where $q = h^2\lambda^2$. Now when $q = 0$ this polynomial has a double root at $+1$. For q slightly greater than zero these roots separate into $\omega_1 < 1$, $\omega_2 > 1$ and there are also two other roots of (3.75a) which we denote by ω_3, ω_4 and which are both larger in modulus than ω_1. It is clear that in order for the solution which we obtain to satisfy the given boundary conditions we need convergence to a root which has modulus less than unity and it turns out that for small q the root which we require is ω_1. It follows immediately from Lemma 3.2 that we obtain convergence to a decreasing solution if the root ω_i of the smallest modulus of (3.75a) is such that $|\omega_i| < 1$. This prompts us to give the following definition of the stability of scheme (3.74) used in conjunction with algorithm (2.5).

Definition 3.5
Consider the solution of the scalar test equation $y'' = \lambda^2 y$ using the finite difference scheme (3.74) where the solution is generated using algorithm (2.5). This procedure is said to be absolutely stable for a given value of $q = h^2\lambda^2$ if, for that value of q, the root ω_1 of smallest modulus of (3.75a) satisfies $|\omega_1| < 1$ and all other roots ω_i satisfy $|\omega_i| \neq |\omega_1|$.

An interval I of the positive real axis is said to be an interval of absolute stability of this procedure if the procedure is absolutely stable for all $q \in I$. Note that we are no longer referring to the stability properties of a discretisation algorithm alone, but rather to the stability of a complete procedure (i.e. discretisation algorithm + precise method of generating the required solution). The interval of absolute stability of scheme (3.74) used in conjunction with algorithm (2.5) may easily be found using a numerical approach and may be shown to be $(0, \infty)$.

Our composite algorithm for the numerical solution of the linear equation (3.70) with boundary conditions (3.71b) has now been fully described and the only remaining question concerns the rate of convergence of our algorithm as a function of N. Practical experience has shown that in general the smaller the value of q the more slowly our algorithm converges. In Table 3.7 we present the results obtained for the numerical integration of the test equation $y'' = \lambda^2 y$ for various values of N and for a range of values of q. The initial condition $y(0) = 1$ is used and the problem considered is to find the first 20 values of y. As can be seen the algorithm is rather slowly convergent for $q = 0{\cdot}1$ but the rate of convergence increases as q increases. In view of this a main aim of future research is to find ways to improve this convergence rate. As mentioned previously the main advantage of this algorithm is that it *does provide us with a practical approximation to infinity* and we shall demonstrate this feature by means of the following test problem involving a parabolic equation:

$$\frac{\partial u}{\partial t} = \frac{\partial^2 u}{\partial x^2} - 2u, \qquad x \geq 0, t \geq 0,$$

TABLE 3.7

(1) $q = 0{\cdot}1$

x	$N = 22$ Solution obtained	$N = 28$ Solution obtained	$N = 34$ Solution obtained	$N = 40$ Solution obtained	True solution
0·1	0·90179009	0·90393173	0·90456797	0·90475882	0·90483742
0·2	0·81260486	0·81690960	0·81818846	0·81857205	0·81873075
0·3	0·73155178	0·73806273	0·73999700	0·74057719	0·74081822
0·4	0·65781971	0·66660204	0·66921109	0·66999369	0·67032005
0·5	0·59067077	0·60181240	0·60512235	0·60611519	0·60653066
1·0	0·33209842	0·35722598	0·36469091	0·36693006	0·36787944
2·0	0·027131062	0·10242522	0·12546069	0·13237136	0·13533528

(2) $q = 0{\cdot}3$

x	$N = 22$ Solution obtained	$N = 28$ Solution obtained	$N = 34$ Solution obtained	$N = 40$ Solution obtained	True solution
0·1	0·74102452	0·74102653	0·74102659	0·74102659	0·74081822
0·2	0·54911608	0·54912029	0·54912040	0·54912040	0·54881164
0·3	0·40690584	0·40691263	0·40691282	0·40691282	0·40656966
0·4	0·30152296	0·30153294	0·30153321	0·30153322	0·30119421
0·5	0·22342965	0·22344374	0·22344412	0·22344413	0·22313016
1·0	0·049858994	0·049925416	0·049927230	0·049927272	0·049787068
2·0	0·0012134157	0·0024550780	0·0024917042	0·0024925633	0·0024787522

(3) $q = 0{\cdot}5$

x	$N = 22$ Solution obtained	$N = 28$ Solution obtained	$N = 34$ Solution obtained	True solution
0·1	0·60770053	0·60770053	0·60770053	0·60653066
0·2	0·36929994	0·36929994	0·36829994	0·36787944
0·3	0·22442377	0·22442377	0·22442377	0·22313016
0·4	0·13638244	0·13638244	0·13638244	0·13533528
0·5	0·082879674	0·082879684	0·082879684	0·082084999
1·0	0·0068689242	0·0068690417	0·0068690420	0·0067379470
2·0	$0{\cdot}31733027 \times 10^{-4}$	$0{\cdot}4713946 \times 10^{-4}$	$0{\cdot}47181528 \times 10^{-4}$	$0{\cdot}45399930 \times 10^{-4}$

with boundary conditions

$$u(x,0) = e^{-(x+2)},$$

$$u(0,t) = e^{-(t+2)},$$

$$u(x,t) \to 0 \quad \text{as} \quad x \to \infty \quad \text{for all} \quad t \geq 0.$$

We generate the solution of this problem using a *semi-discretisation approach.* By making the finite difference approximation

$$\frac{\partial u\left(x, t+\dfrac{\delta t}{2}\right)}{\partial t} = \frac{u(x, t+\delta t) - u(x,t)}{\delta t}$$

we have that

$$\frac{d^2 u\left(x, t+\dfrac{\delta t}{2}\right)}{dx^2} = \frac{u(x, t+\delta t) - u(x,t)}{\delta t} + 2u\left(x, t+\frac{\delta t}{2}\right).$$

We then set $\omega(x, t + \delta t/2) = u(x, t + \delta t) + u(x, t)$ and make the approximations

$$\frac{d^2 u}{dx^2}\left(x, t+\frac{\delta t}{2}\right) = \frac{1}{2}\frac{d^2\omega}{dx^2}\left(x, t+\frac{\delta t}{2}\right),$$

$$u\left(x, t+\frac{\delta t}{2}\right) = \frac{1}{2}\omega\left(x, t+\frac{\delta t}{2}\right),$$

so that

$$\frac{d^2\omega}{dx^2}\left(x, t+\frac{\delta t}{2}\right) - \frac{2\omega}{\delta t}\left(x, t+\frac{\delta t}{2}\right) = -\frac{4u(x,t)}{\delta t} + 2\omega\left(x, t+\frac{\delta t}{2}\right). \tag{3.75b}$$

We may now integrate this second order ordinary differential equation to obtain the required solution $\omega(x, t + \delta t/2)$ using scheme (3.74). The particular problem considered is that in which we take $\delta t = h = 0{\cdot}1$ and the solution is required with an absolute error not greater than (a) $0{\cdot}5 \times 10^{-2}$, (b) $0{\cdot}5 \times 10^{-3}$ and (c) $0{\cdot}5 \times 10^{-4}$ in the range $0 \leq x \leq 1$. The results obtained for this problem are given in Table 3.8. These results are largely self-explanatory and, as can be seen, the value of N is estimated so that the required degree of precision is achieved over the prescribed range for all t.

We now consider the problem of developing a genuine boundary value approach to the solution of (3.70) in cases where the boundary conditions are

TABLE 3.8a

Estimated value of $N = 16$

	$t = 0{\cdot}1$		$t = 0{\cdot}5$		$t = 1{\cdot}0$	
x	Solution obtained	Absolute error	Solution obtained	Absolute error	Solution obtained	Absolute error
0·1	0·110773	$0{\cdot}305 \times 10^{-4}$	$0{\cdot}742531 \times 10^{-1}$	$0{\cdot}204 \times 10^{-4}$	$0{\cdot}450368 \times 10^{-1}$	$0{\cdot}124 \times 10^{-4}$
0·2	0·100193	$0{\cdot}633 \times 10^{-4}$	$0{\cdot}671611 \times 10^{-1}$	$0{\cdot}444 \times 10^{-4}$	$0{\cdot}407353 \times 10^{-1}$	$0{\cdot}269 \times 10^{-4}$
0·3	$0{\cdot}906025 \times 10^{-1}$	$0{\cdot}115 \times 10^{-3}$	$0{\cdot}607327 \times 10^{-1}$	$0{\cdot}774 \times 10^{-4}$	$0{\cdot}368362 \times 10^{-1}$	$0{\cdot}469 \times 10^{-4}$
0·4	$0{\cdot}818956 \times 10^{-1}$	$0{\cdot}189 \times 10^{-3}$	$0{\cdot}548963 \times 10^{-1}$	$0{\cdot}127 \times 10^{-3}$	$0{\cdot}332963 \times 10^{-1}$	$0{\cdot}770 \times 10^{-4}$
0·5	$0{\cdot}739688 \times 10^{-1}$	$0{\cdot}305 \times 10^{-3}$	$0{\cdot}495828 \times 10^{-1}$	$0{\cdot}204 \times 10^{-3}$	$0{\cdot}300735 \times 10^{-1}$	$0{\cdot}124 \times 10^{-3}$
0·6	$0{\cdot}667167 \times 10^{-1}$	$0{\cdot}489 \times 10^{-3}$	$0{\cdot}447215 \times 10^{-1}$	$0{\cdot}328 \times 10^{-3}$	$0{\cdot}271250 \times 10^{-1}$	$0{\cdot}199 \times 10^{-3}$
0·7	$0{\cdot}600294 \times 10^{-1}$	$0{\cdot}781 \times 10^{-3}$	$0{\cdot}402389 \times 10^{-1}$	$0{\cdot}523 \times 10^{-3}$	$0{\cdot}244061 \times 10^{-1}$	$0{\cdot}317 \times 10^{-3}$
0·8	$0{\cdot}537694 \times 10^{-1}$	$0{\cdot}125 \times 10^{-2}$	$0{\cdot}360427 \times 10^{-1}$	$0{\cdot}840 \times 10^{-3}$	$0{\cdot}218610 \times 10^{-1}$	$0{\cdot}510 \times 10^{-3}$
0·9	$0{\cdot}477897 \times 10^{-1}$	$0{\cdot}200 \times 10^{-2}$	$0{\cdot}320344 \times 10^{-1}$	$0{\cdot}133 \times 10^{-2}$	$0{\cdot}194299 \times 10^{-1}$	$0{\cdot}812 \times 10^{-3}$
1·0	$0{\cdot}418236 \times 10^{-1}$	$0{\cdot}323 \times 10^{-2}$	$0{\cdot}280352 \times 10^{-1}$	$0{\cdot}216 \times 10^{-2}$	$0{\cdot}170042 \times 10^{-1}$	$0{\cdot}131 \times 10^{-2}$

TABLE 3.8b

Estimated value of $N = 25$

	$t = 0{\cdot}1$		$t = 0{\cdot}5$		$t = 1{\cdot}0$	
x	Solution obtained	Absolute error	Solution obtained	Absolute error	Solution obtained	Absolute error
0·1	0·110800	$0{\cdot}286 \times 10^{-5}$	$0{\cdot}742717 \times 10^{-1}$	$0{\cdot}192 \times 10^{-5}$	$0{\cdot}450480 \times 10^{-1}$	$0{\cdot}116 \times 10^{-5}$
0·2	0·100254	$0{\cdot}459 \times 10^{-5}$	$0{\cdot}672024 \times 10^{-1}$	$0{\cdot}307 \times 10^{-5}$	$0{\cdot}407603 \times 10^{-1}$	$0{\cdot}186 \times 10^{-5}$
0·3	$0{\cdot}907122 \times 10^{-1}$	$0{\cdot}572 \times 10^{-5}$	$0{\cdot}608062 \times 10^{-1}$	$0{\cdot}384 \times 10^{-5}$	$0{\cdot}368808 \times 10^{-1}$	$0{\cdot}233 \times 10^{-5}$
0·4	$0{\cdot}820783 \times 10^{-1}$	$0{\cdot}668 \times 10^{-5}$	$0{\cdot}550187 \times 10^{-1}$	$0{\cdot}448 \times 10^{-5}$	$0{\cdot}333706 \times 10^{-1}$	$0{\cdot}272 \times 10^{-5}$
0·5	$0{\cdot}742658 \times 10^{-1}$	$0{\cdot}781 \times 10^{-5}$	$0{\cdot}497818 \times 10^{-1}$	$0{\cdot}524 \times 10^{-5}$	$0{\cdot}301942 \times 10^{-1}$	$0{\cdot}318 \times 10^{-5}$
0·6	$0{\cdot}671960 \times 10^{-1}$	$0{\cdot}951 \times 10^{-5}$	$0{\cdot}450428 \times 10^{-1}$	$0{\cdot}637 \times 10^{-5}$	$0{\cdot}273199 \times 10^{-1}$	$0{\cdot}387 \times 10^{-5}$
0·7	$0{\cdot}607978 \times 10^{-1}$	$0{\cdot}123 \times 10^{-4}$	$0{\cdot}407540 \times 10^{-1}$	$0{\cdot}822 \times 10^{-5}$	$0{\cdot}247185 \times 10^{-1}$	$0{\cdot}498 \times 10^{-5}$
0·8	$0{\cdot}550064 \times 10^{-1}$	$0{\cdot}168 \times 10^{-4}$	$0{\cdot}368719 \times 10^{-1}$	$0{\cdot}113 \times 10^{-4}$	$0{\cdot}223639 \times 10^{-1}$	$0{\cdot}683 \times 10^{-5}$
0·9	$0{\cdot}497628 \times 10^{-1}$	$0{\cdot}242 \times 10^{-4}$	$0{\cdot}333570 \times 10^{-1}$	$0{\cdot}162 \times 10^{-4}$	$0{\cdot}202321 \times 10^{-1}$	$0{\cdot}985 \times 10^{-5}$
1·0	$0{\cdot}450128 \times 10^{-1}$	$0{\cdot}364 \times 10^{-4}$	$0{\cdot}301730 \times 10^{-1}$	$0{\cdot}244 \times 10^{-4}$	$0{\cdot}183009 \times 10^{-1}$	$0{\cdot}148 \times 10^{-4}$

TABLE 3.8c

Estimated value of $N = 31$

	$t = 0{\cdot}1$		$t = 0{\cdot}5$		$t = 1{\cdot}0$	
x	Solution obtained	Absolute error	Solution obtained	Absolute error	Solution obtained	Absolute error
0·1	0·110801	$0{\cdot}259 \times 10^{-5}$	$0{\cdot}742718 \times 10^{-1}$	$0{\cdot}173 \times 10^{-5}$	$0{\cdot}450482 \times 10^{-1}$	$0{\cdot}105 \times 10^{-5}$
0·2	0·100255	$0{\cdot}397 \times 10^{-5}$	$0{\cdot}672029 \times 10^{-1}$	$0{\cdot}266 \times 10^{-5}$	$0{\cdot}407606 \times 10^{-1}$	$0{\cdot}161 \times 10^{-5}$
0·3	$0{\cdot}907133 \times 10^{-1}$	$0{\cdot}462 \times 10^{-5}$	$0{\cdot}608070 \times 10^{-1}$	$0{\cdot}310 \times 10^{-5}$	$0{\cdot}368813 \times 10^{-1}$	$0{\cdot}188 \times 10^{-5}$
0·4	$0{\cdot}820802 \times 10^{-1}$	$0{\cdot}484 \times 10^{-5}$	$0{\cdot}550200 \times 10^{-1}$	$0{\cdot}324 \times 10^{-5}$	$0{\cdot}333713 \times 10^{-1}$	$0{\cdot}197 \times 10^{-5}$
0·5	$0{\cdot}742688 \times 10^{-1}$	$0{\cdot}482 \times 10^{-5}$	$0{\cdot}497838 \times 10^{-1}$	$0{\cdot}323 \times 10^{-5}$	$0{\cdot}301954 \times 10^{-1}$	$0{\cdot}196 \times 10^{-5}$
0·6	$0{\cdot}672008 \times 10^{-1}$	$0{\cdot}468 \times 10^{-5}$	$0{\cdot}450461 \times 10^{-1}$	$0{\cdot}314 \times 10^{-5}$	$0{\cdot}273218 \times 10^{-1}$	$0{\cdot}190 \times 10^{-5}$
0·7	$0{\cdot}608056 \times 10^{-1}$	$0{\cdot}451 \times 10^{-5}$	$0{\cdot}407592 \times 10^{-1}$	$0{\cdot}302 \times 10^{-5}$	$0{\cdot}247217 \times 10^{-1}$	$0{\cdot}183 \times 10^{-5}$
0·8	$0{\cdot}550189 \times 10^{-1}$	$0{\cdot}436 \times 10^{-5}$	$0{\cdot}368802 \times 10^{-1}$	$0{\cdot}292 \times 10^{-5}$	$0{\cdot}223690 \times 10^{-1}$	$0{\cdot}177 \times 10^{-5}$
0·9	$0{\cdot}497828 \times 10^{-1}$	$0{\cdot}429 \times 10^{-5}$	$0{\cdot}333704 \times 10^{-1}$	$0{\cdot}288 \times 10^{-5}$	$0{\cdot}202402 \times 10^{-1}$	$0{\cdot}175 \times 10^{-5}$
1·0	$0{\cdot}450448 \times 10^{-1}$	$0{\cdot}439 \times 10^{-5}$	$0{\cdot}301944 \times 10^{-1}$	$0{\cdot}294 \times 10^{-5}$	$0{\cdot}183139 \times 10^{-1}$	$0{\cdot}178 \times 10^{-5}$

of the form (3.71a). This is most conveniently done by considering an example in which we have a discretisation scheme with the general form

$$y_{n+2} - 2y_{n+1} + y_n = h^2\{py''_{n+4} + qy''_{n+3} + sy''_{n+2} + ty''_{n+1} + uy''_n\}. \quad (3.75c)$$

It may easily be verified that the coefficients in (3.75c) can be chosen so that (3.75c) has order 6 and the corresponding scheme is given by

$$y_{n+2} - 2y_{n+1} + y_n$$
$$= h^2\{-\tfrac{1}{240}y''_{n+4} + \tfrac{1}{60}y''_{n+3} + \tfrac{7}{120}y''_{n+2} + \tfrac{204}{240}y''_{n+1} + \tfrac{19}{240}y''_n\}. \quad (3.76)$$

If we now apply this numerical integration procedure to the solution of (3.70) we obtain a fourth order linear recurrence relation of the form

$$\sum_{i=0}^{4} a_i(n)y_{n+i} = 0, \quad (3.77)$$

where the $a_i(n)$ are defined uniquely in terms of $f(x_n)$ and the coefficients appearing in equation (3.76). We seek to extract from (3.77) a "quadratic factor" of the form

$$y_{n+2} = r(n)y_{n+1} + s(n)y_n \quad (3.78)$$

using the approach described in Section 2.6. Having done this we rewrite equations (3.78) as a tri-diagonal system of algebraic equations and use the given boundary conditions to solve for y_n in the usual way. Thus the first stage of our algorithm does not involve the computation of y_n, but instead "reduces" (3.76) to a second order linear recurrence relation the solutions of which form a basis for the solutions of (3.70). We may illustrate this approach most conveniently by considering the case $f(x_n) \equiv \lambda^2$ = a real constant. In this case (3.77) reduces to the constant coefficient equation

$$\sum_{i=0}^{4} a_i y_{n+i} = 0$$

the general solution of which takes the form

$$y_n = \sum_{i=1}^{4} c_i \omega_i^n,$$

where the ω_i are the four roots, assumed distinct, of the polynomial $\Sigma_{i=0}^4 a_i x^i = 0$. In the limit $q \equiv h\lambda = 0$ two of the ω_i are unity (the principal roots) and two of the ω_i are zero (spurious roots). As q increases the ω_i become distinct but for a sufficiently small region I, at least, of the positive real axis the roots ω_1, ω_2 will still be clearly defined. If we were now to apply the first stage of our approach to the solution of this simple test problem for some value of $q \in I$ our aim would be to derive a quadratic factor of the form

$$y_{n+2} = ry_{n+1} + sy_n$$

which has as its solutions ω_1, ω_2; i.e. the principal roots of our polynomial. In this way the spurious roots of our original recurrence relation are removed from the numerical solution which we actually obtain. It is clear that in order that this approach should be applicable it is necessary that the principal roots of our quartic polynomial should be well defined. By taking a range of values of the real quantity λ these roots may be computed explicitly for any given λ. It is found that the two principal roots remain well defined for all real λ with the spurious roots always being complex with relatively large modulus.

In order to extract the quadratic factor (3.78) from (3.77) in the more general case where $f(x_n)$ is no longer assumed to be a constant it is necessary to derive the algebraic equations satisfied by $r(n)$ and $s(n)$. Following the approach developed in Section 2.6 we do this by substituting (3.78) into (3.77) to produce a first order linear recurrence relation of the form $A_n y_{n+1} = B_n y_n$ for particular values of the functions A_n and B_n. Since this relation is valid for all solutions of (3.78) it follows immediately that $A_n = B_n = 0$ which in turn gives the relations

$$\left.\begin{aligned} &a_4(n)\{r(n+2)r(n+1)r(n) + r(n+2)s(n+1) + s(n+2)r(n)\} \\ &\qquad + a_3(n)\{r(n+1)r(n) + s(n+1)\} + a_2(n)r(n) + a_1(n) = 0, \\ &a_4(n)\{r(n+2)r(n+1)s(n) + s(n+2)s(n)\} \\ &\qquad + a_3(n)r(n+1)s(n) + a_2(n)s(n) + a_0(n) = 0. \end{aligned}\right\} \tag{3.79}$$

An examination of our finite difference scheme (3.76) reveals that both of the coefficients $a_4(n)$ and $a_3(n)$, as defined in (3.77), are small for small values of $h^2 y_i''$, (and usually h will be quite small in order to ensure a reasonable degree of accuracy). We would thus expect reasonably rapid convergence to occur if we were to use the iteration scheme

$$\left.\begin{aligned} &a_4(n)\{r^{(p-1)}(n+2)r^{(p-1)}(n+1)r^{(p)}(n) \\ &\qquad + r^{(p-1)}(n+2)s^{(p-1)}(n+1) + s^{(p-1)}(n+2)r^{(p)}(n)\} \\ &\qquad + a_3(n)\{r^{(p-1)}(n+1)r^{(p)}(n) + s^{(p-1)}(n+1)\} \\ &\qquad + a_2(n)r^{(p)}(n) + a_1(n) = 0, \\ &a_4(n)\{r^{(p-1)}(n+2)r^{(p-1)}(n+1)s^{(p)}(n) + s^{(p-1)}(n+2)s^{(p)}(n)\} \\ &\qquad + a_3(n)r^{(p-1)}(n+1)s^{(p)}(n) + a_2(n)s^{(p)}(n) + a_0(n) = 0. \end{aligned}\right\} \tag{3.80}$$

Practical experience has shown that this iteration scheme usually does converge fairly rapidly to the two required sequences $r(n)$ and $s(n)$, providing the product $h^2 y_i''$ is "reasonably small", but no theoretical justification of this process is available at the present time. In order to use this iteration scheme we need initial sequences $r^{(0)}(n)$ and $s^{(0)}(n)$ to be available and, in the absence of any additional information, we usually use $r^{(0)}(n) \equiv s^{(0)}(n) = 0$ for all n. Having derived this quadratic factor, assuming of course that this iteration converges, the first stage of our algorithm is complete. However, before going on to consider the second stage we first consider another method which was described in Section 2.6 for determining the sequences $r(n)$ and $s(n)$. In this approach we set $r(\bar{N}+1) = r(\bar{N}+2) = s(\bar{N}+1) = s(\bar{N}+2) = 0$ for some large value of $\bar{N} \geq N$ and then, substituting into (3.79) for $n = \bar{N}$, we obtain

$$r(\bar{N}) = -\frac{a_1(\bar{N})}{a_2(\bar{N})}, \qquad s(\bar{N}) = -\frac{a_0(\bar{N})}{a_2(\bar{N})}.$$

Once these two values have been obtained, we may solve equations (3.79) for the two sequences $r(n)$ and $s(n)$ for all $n \doteq \bar{N}-1, \bar{N}-2, \ldots, 0$; i.e. in the direction of decreasing n. We note that at each stage it is not necessary to solve a pair of simultaneous equations for these sequences, since $r(n)$ is given immediately from the first of (3.80) and $s(n)$ is given from the second of (3.80). This procedure is still not a completely satisfactory extension of Olver's technique, since it is necessary to guess a value of $\bar{N}$ at which to set $r(\bar{N}+1) = s(\bar{N}+1) = 0$ initially. The procedure which we shall in fact use in practice is very similar to that first used by Miller (1952) in conjunction with his backward recurrence algorithm. We denote the two sequences $r(n)$ and $s(n)$ obtained as the solution of (3.80) when we set $r(\bar{N}+1) = s(\bar{N}+1) = 0$ by $r^{[\bar{N}]}(n)$ and $s^{[\bar{N}]}(n)$. Having obtained these two sequences we then compute the two additional sequences $r^{[\bar{N}+5]}(n)$ and $s^{[\bar{N}+5]}(n)$ for all $0 \leq n \leq \bar{N}+5$. If the two sequences of approximations to $r(n)$ and the two sequences of approximations to $s(n)$ are both sufficiently close in some appropriate sense, the two sequences $r^{[\bar{N}+5]}(n)$ and $s^{[\bar{N}+5]}(n)$ are taken as our finally accepted approximations. If this is not so the same procedure is carried out again starting from the point $\bar{N}+10$. This initial need to estimate a suitable value of $\bar{N}$ is not as serious in genuine boundary value cases as it is for purely initial value problems, since the value $\bar{N} = N + 5$ is normally a good starting point for our calculations. We again emphasise that this procedure is based mainly on numerical evidence and there is still a need for some useful analytic results regarding the convergence of this process.

Having extracted a quadratic factor of the form (3.78) from our original finite difference equation we are left with the problem of solving a system of algebraic equations of the form

$$\mathbf{Ay} = \mathbf{b}, \qquad y(x_0) = y_0, \qquad y(x_N) = y_N,$$

where the matrix **A** is tri-diagonal and **b** is a known vector. This may be done very efficiently by standard methods (see Richtmeyer and Morton, 1967, p. 200 for example). We will assume that this process is stable.

We now illustrate this second approach by considering a particular numerical example. We adhere to our usual procedure of considering particularly simple problems, since it is generally rather more easy to see exactly what is happening in these cases. The particular problem under consideration is that in which

$$y'' = y, \qquad y(0) = 1, \qquad y(1) = \exp(-1),$$

using a steplength of $h = 0{\cdot}1$. The solution of this problem, which is given by $y(x) = \exp(-x)$, is required at the step points $0{\cdot}1(0{\cdot}1)0{\cdot}9$. Applying (3.76) to solve this equation we obtain the recurrence relation

$$\frac{y_{n+4}}{24000} - \frac{y_{n+3}}{6000} + \left(1 - \frac{7}{12000}\right)y_{n+2} - \left(2 + \frac{17}{2000}\right)y_{n+1} + \left(1 - \frac{19}{24000}\right)y_n = 0. \tag{3.81}$$

The general solution of this recurrence relation takes the form

$$y_n = \sum_{i=1}^{4} c_i \omega_i^n,$$

where the c_i are constants and to eight significant figures $\omega_1 = 0{\cdot}90483742$, $\omega_2 = 1{\cdot}1051709$, $\omega_3, \omega_4 = 0{\cdot}99499583 \pm 154{\cdot}85480i$. It is clear that our principal roots are ω_1, ω_2 and the spurious roots are ω_3, ω_4. The first part of our algorithm is to extract from (3.81) a quadratic factor of the form

$$y_{n+2} = ry_{n+1} + sy_n, \tag{3.82}$$

which has a general solution of the form

$$y_n = \sum_{i=1}^{2} c_i \omega_i^n;$$

i.e. which does not contain components of the spurious solutions ω_3 and ω_4. By taking the values $\bar{N} = 15, 20$ it is found that the two sets of values for r and the two sets of values for s are both in agreement to at least ten significant figures over the range $1 \leq n \leq 9$ and the values actually obtained are

$$r = 2{\cdot}010008332, \qquad s = -0{\cdot}9999999958.$$

The solutions of this quadratic factor are clearly linear combinations of ω_1 and ω_2 (as required) and so do not contain components of ω_3 or ω_4. Having obtained this quadratic factor we use the given boundary conditions to produce the required solution in the usual way and the solution actually obtained is given in Table 3.9. The way that this approach can be extended

TABLE 3.9

n	Solution obtained	True solution
1	0·9048374169	0·9048374180
2	0·8187307511	0·8187307531
3	0·7408182182	0·7408182207
4	0·6703200433	0·6703200460
5	0·6065306570	0·6065306597
6	0·5488116335	0·5488116361
7	0·4965853016	0·4965853038
8	0·4493289625	0·4493289641
9	0·4065696589	0·4065696597

should be clear and this extension is now outlined. The first stage of our algorithm is to produce a discretisation scheme of the form

$$y_{n+2} - 2y_{n+1} + y_n = h^2 \sum_{i=0}^{t} \alpha_i y''_{n+i} \tag{3.83}$$

suitable for the numerical solution of (3.70). It is worthwhile pointing out at this stage that scheme (3.76) may be written in the form

$$\underbrace{y_{n+2} - 2y_{n+1} + y_n = \frac{h^2}{12}\{y''_{n+2} + 10y''_{n+1} + y''_n\}}_{\text{usual fourth order method}} - \frac{1}{240}\Delta^4 y''_n. \tag{3.84}$$

In view of this relation, scheme (3.76) may be regarded as arising from the use of Numerov's method (the bracketed part) together with a first correction term. By applying still higher order correction terms we may obtain increasingly accurate formulae of the type (3.83) and it is clear that this approach has connections with the method of deferred correction developed by Pereyra (1967). Although this is a convenient way of deriving integration procedures it is, of course, by no means the only way of deriving higher order schemes of the form (3.83). If we now apply (3.83) to the solution of the scalar equation $y'' = \lambda^2 y$, we obtain a linear recurrence relation with constant coefficients of the general form

$$\sum_{i=0}^{t} a_i y_{n+i} = 0.$$

The general solution of this recurrence relation is

$$y_n = \sum_{i=1}^{t} c_i \omega_i^n,$$

where the ω_i are the roots, assumed distinct, of the polynomial equation

$$\sum_{i=0}^{t} a_i x^i = 0.$$

In the limit $q = 0$ this polynomial has a double root at $+1$ and a root at 0 of multiplicity $t-2$. Using the nomenclature adopted earlier we will call the two roots at $+1$ the *principal roots* and we will refer to the others as the *spurious roots*. As q increases these ω_i will become distinct but, as before, the two principal roots will still be clearly defined for sufficiently small q. The next stage of our algorithm is to extract from (3.83) a quadratic factor of the form (3.78) so that the two solutions of (3.78) correspond to the principal roots of (3.83) in the linear case. This may be done by substituting (3.78) into (3.83) to produce a first order relation of the form $A_n y_{n+1} = B_n y_n$. Setting $A_n = B_n = 0$ we obtain two non-linear algebraic equations for $r(n)$ and $s(n)$ and these equations may be solved using a natural extension of (3.80) or by using an extension of the backward recurrence algorithm described earlier. Having obtained our quadratic factor we solve for the required solution in exactly the same way as previously described. We again emphasise that our procedures have evolved mainly from practical experience and are not at present backed up by theoretical results. In particular, results regarding the convergence of both algorithm (3.80) and the algorithm based on backward recurrence to limit sequences $r(n)$ and $s(n)$ are needed as are theoretical results which tell us under what circumstances the quadratic factor produced by these algorithms (in cases where they do converge) is the required quadratic factor of (3.83).

We may further tentatively extend this approach to deal with the numerical solution of a certain class of non-linear boundary value problems. In order that we may be able to do this it is first of all necessary for us to extend slightly some of the techniques developed earlier by deriving a procedure for extracting a quadratic factor of the form

$$y_{n+2} = r(n)y_{n+1} + s(n)y_n + g(n) \tag{3.85}$$

from the fourth order linear, inhomogeneous recurrence relation

$$\sum_{i=0}^{4} a_i(n)y_{n+4} = f(n). \tag{3.86}$$

By following through a very similar analysis to that considered earlier for the homogeneous case we find that the sequences $r(n)$, $s(n)$ and $g(n)$ appearing in (3.86) satisfy the relations

$$\left.\begin{aligned}
&a_4(n)[r(n+2)r(n+1)r(n) + r(n+2)s(n+1) + s(n+2)r(n)] \\
&\qquad + a_3(n)[r(n+1)r(n) + s(n+1)] + a_2(n)r(n) + a_1(n) = 0, \\
&a_4(n)[r(n+2)r(n+1)s(n) + s(n+2)s(n)] \\
&\qquad + a_3(n)r(n+1)s(n) + a_2(n)s(n) + a_0(n) = 0, \\
&a_4(n)[r(n+2)r(n+1)g(n) + r(n+2)g(n+1) \\
&\qquad + s(n+2)g(n) + g(n+2)] \\
&\qquad + a_3(n)[r(n+1)g(n) + g(n+1)] + a_2(n)g(n) - f(n) = 0,
\end{aligned}\right\} \tag{3.87}$$

and a solution of this non-linear system of algebraic equations may be generated over the range of interest by using either of the two techniques discussed earlier in this section.

We now examine how this technique may be used to generate the required solution of a certain class of non-linear boundary value problems of the form

$$y'' = f(x, y), \qquad y(x_0) = y_0, \qquad y(x_N) = y_N. \tag{3.88}$$

Applying (3.76) to the solution of this problem we obtain the non-linear recurrence relation

$$y_{n+2} - 2y_{n+1} + y_n = h^2\{-\tfrac{1}{240}f_{n+4} + \tfrac{1}{60}f_{n+3} + \tfrac{7}{120}f_{n+2} + \tfrac{17}{20}f_{n+1} + \tfrac{19}{240}f_n\}. \tag{3.89}$$

In order that we may be able to apply the techniques developed earlier in this section to the solution of (3.89) it is necessary for us to first of all linearise our problem. We do this by making use of the approximation

$$f_i^{(p)} = f_i^{(p-1)} + [y_i^{(p)} - y_i^{(p-1)}]\frac{\partial f}{\partial y}(x_i, y_i^{(p-1)}), \tag{3.90}$$

where the $y_i^{(p)}$ satisfy the relation

$$y_{n+2}^{(p)} - 2y_{n+1}^{(p)} + y_n^{(p)} = h^2\{-\tfrac{1}{240}f_{n+4}^{(p)} + \tfrac{1}{60}f_{n+3}^{(p)} + \tfrac{7}{120}f_{n+2}^{(p)} + \tfrac{17}{20}f_{n+1}^{(p)} + \tfrac{19}{240}f_n^{(p)}\}. \quad (3.91)$$

In order to be able to use this approach we first of all need a sequence of initial approximations $\{y_n^{(0)}\}$ to $\{y_n\}$. Assuming for the time being that such a sequence is available and substituting (3.90) into (3.91) for $p = 1$ we obtain a recurrence relation of the form

$$\sum_{i=0}^{4} a_i(n)y_{n+i}^{(1)} = e(n)$$

for appropriate values of the sequences $a_i(n)$ and $e(n)$. We may now extract a quadratic factor from this recurrence relation and then solve for the required solution $\{y_n^{(1)}\}$ using the techniques described earlier in this section. Having obtained this sequence of approximations $\{y_n^{(1)}\}$ we may substitute into (3.90) for $p = 2$ to obtain an approximation for $f_i^{(2)}$. We then substitute this approximation into (3.91) for $p = 2$ to give the recurrence relation satisfied by the sequence $\{y_n^{(2)}\}$. We may once again go through the procedure of extracting a quadratic factor from this recurrence relation and then solving for the required sequence of approximations in the way previously described. By carrying out this procedure we obtain sequences of approximations $\{y_n^{(p)}\}$, $p = 1, 2, \ldots$, which hopefully will converge to the required solution $y(x_n)$ with increasing p in the sense that

$$\lim_{p\to\infty} \|y(x_n) - y_n^{(p)}\| = 0.$$

This procedure, depending as it does on a Newton-type approximation, is rather hard to analyse and once again we have to rely mainly on numerical evidence to support the claim that it can be useful in practice. One factor which will crucially affect the convergence properties of this scheme is the accuracy of the initial approximations $\{y_n^{(0)}\}$, since it is often the case that these have to be reasonably accurate in order that the Newton iteration schemes should converge at all. In view of this, one possible way of utilising our scheme is to use it rather like a corrector in a predictor–corrector scheme. This may most conveniently be done in the following manner. A solution of our problem is first generated using the straightforward Numerov scheme

$$y_{n+2} - 2y_{n+1} + y_n = \frac{h^2}{12}\{f_{n+2} + 10f_{n+1} + f_n\}.$$

The solution y_n generated by this scheme, which has a local truncation error of order $0(h^6)$, may be used as the first approximation, $y_n^{(0)}$, in the iteration scheme which we have just derived and which is of higher order with a local truncation error of order $0(h^7)$. It is found that, providing h is "sufficiently small" and the function f is "slowly varying", this scheme generally does have good convergence characteristics. In general, however, the non-linear case can present severe convergence problems (as it also can do with more conventional boundary value techniques) and this whole approach must be applied with care in this particular case.

A slightly different scheme which involves less computational effort but which does not generally have such good convergence characteristics as the one just described may be derived in the following way. In order to generate a solution of (3.89) we use the iteration scheme

$$\begin{aligned} &y_{n+2}^{(t)} - 2y_{n+1}^{(t)} + y_n^{(t)} \\ &\quad = h^2\{-\tfrac{1}{240}f_{n+4}^{(t-1)} + \tfrac{1}{60}f_{n+3}^{(t-1)} + \tfrac{7}{120}f_{n+2}^{(t)} + \tfrac{17}{20}f_{n+1}^{(t)} + \tfrac{19}{240}f_n^{(t)}\}. \end{aligned} \tag{3.92}$$

If we now use approximation (3.90) in (3.92), we immediately obtain a quadratic factor of the form (3.78) and we may then solve for the required solution in the usual way. This scheme may also be used as the corrector in a predictor–corrector scheme with a more simple lower order scheme being used to generate $y_n^{(0)}$ in the way previously described. One added difficulty associated with both of these approaches is that in order to compute our sequence of iterates it is necessary to compute values of y outside the far end of the range of integration. By following the schemes through in detail it can soon be seen that the extra values of the solution which are required are y_{N+1} and y_{N+2} and so we have to be sure that no additional problems will arise in the computation of these quantities; e.g. if there is a singularity at the far end of the range of integration. This problem is, however, easy to overcome when using the second approach since halfway through the range of integration we can switch to using the scheme

$$y_{n+2}^{(p)} - 2y_{n+1}^{(p)} + y_n^{(p)} = h^2\{\tfrac{19}{240}f_{n+2}^{(p)} + \tfrac{17}{20}f_{n+1}^{(p)} + \tfrac{7}{120}f_n^{(p)} + \tfrac{1}{60}f_{n-1}^{(p)} - \tfrac{1}{240}f_{n-2}^{(p)}\}.$$

When using the first approach we compute the coefficients of our quadratic factor up to and including $r(N)$, $s(N)$ so that in order to compute our extra solution values we may use (3.78) for $n = N - 1, N$.

We now illustrate the quadratic factor approach and the approach typified by (3.92) by considering a particular example which, as usual, is a fairly simple one. The problem which we consider is given by

$$y'' \doteq 2y^3, \qquad y(0) = 1, \qquad y(1) = \tfrac{1}{2},$$

where the solution is required at the points 0·1(0·1)0·9. The results obtained using the quadratic factor approach are given in Table 3.10a. The initial approximation used is $y_n^{(0)} \equiv 1$ and it is found that convergence to eight decimal places is obtained between the fourth and fifth iterates. As can be seen, convergence is reasonably rapid even though the sequence of initial approximations is not particularly accurate. In Table 3.10b we list the results obtained using the second approach which in general requires less computational effort since there is no need first to extract a quadratic factor. The initial approximation $y_n^{(0)} \equiv 1$ is again used and as can be seen from the tables the rate of convergence is about the same in both cases.

Finally in this section we need to emphasise that the algorithms proposed have been considered in order to show how the ideas formulated in Chapter 2

TABLE 3.10a

x	1st Iterate	2nd Iterate	3rd Iterate	4th Iterate	True solution
0·1	0·91916903	0·90923669	0·90808997	0·90908994	0·90909091
0·2	0·85356417	0·83362641	0·83333208	0·83333320	0·83333333
0·3	0·79922939	0·76966281	0·76922950	0·76922942	0·76923077
0·4	0·75288826	0·71483055	0·71428456	0·71428447	0·71428571
0·5	0·71174638	0·66727701	0·66666656	0·66666558	0·66666667
0·6	0·67332287	0·62561033	0·62499921	0·62499911	0·62500000
0·7	0·63530077	0·58877148	0·58823470	0·58823462	0·58823529
0·8	0·59538732	0·55594967	0·55555516	0·55555510	0·55555556
0·9	0·55117572	0·52652159	0·52631559	0·52631556	0·52631379

TABLE 3.10b

x	1st Iterate	2nd Iterate	3rd Iterate	4th Iterate
0·1	0·91854308	0·90917770	0·90908953	0·90908994
0·2	0·85241531	0·83351908	0·83333132	0·83333320
0·3	0·79766564	0·76951999	0·76922853	0·76922942
0·4	0·75102446	0·71466744	0·71428350	0·71428446
0·5	0·70970843	0·66711049	0·66666465	0·66666558
0·6	0·67125428	0·62545746	0·62499831	0·62499911
0·7	0·63337202	0·58864740	0·58823400	0·58823461
0·8	0·59380852	0·55586516	0·55555469	0·55555509
0·9	0·55021352	0·52648090	0·52631536	0·52631555

may be used in the numerical solution of boundary value problems. We are not claiming that our algorithms are, at the present stage, superior to any existing ones, such as that developed by Pereyra, but we have aimed to show that this approach does seem to be sufficiently promising to merit further investigation.

3.13 Applications to partial differential equations

Finally in this chapter we mention the possible extension of some of the techniques of the previous sections to deal with the approximate numerical solution of the simple *heat conduction equation* with non-derivative boundary conditions.

The particular problem is that given by

$$\frac{\partial u}{\partial t} = \sigma(x,t)\frac{\partial^2 u}{\partial x^2} \tag{3.93}$$

with associated boundary conditions

$$\begin{aligned} u(x,0) &= f(x), \\ u(0,t) &= g_1(t), \\ u(1,t) &= g_2(t), \end{aligned} \tag{3.94}$$

where we assume that the function $\sigma(x, t)$ is a twice continuously differentiable function of both the independent variables x and t. We shall seek to generate the required solution of (3.93) on a discrete set of points with grid spacing Δx in the space direction and Δt in the t direction. We first of all replace the space derivative $\partial^2 u(x,t)/\partial x^2$ by the finite difference approximation

$$\frac{\partial^2 u_{j,n}}{\partial x^2} \equiv \frac{u_{j+1,n} - 2u_{j,n} + u_{j-1,n}}{(\Delta x)^2} - \frac{1}{12}\delta^2\left\{\frac{1}{\sigma(x,t)}\frac{\partial u_{j,n}}{\partial t}\right\}, \tag{3.95}$$

where $x = j\Delta x$, $t = n\Delta t$ and δ^2 denotes the usual central difference operator. It may be shown that this approximation has local truncation error of order $0(\Delta x)^4$. Following the procedure adopted in some of the earlier sections of this chapter we replace the time derivative by a scheme involving values of the required solution at more than one advanced time step (since this allows us to achieve relatively high orders of accuracy). We then solve the resulting finite difference equation iteratively in a precisely defined way to yield a stable, rapidly convergent method. For the purposes of our demonstration we replace the time derivative $\partial u(x,t)/\partial t$ by the approximation

$$\frac{\partial u_{j,n}}{\partial t} = \frac{u_{j,n+3} - \frac{17}{4}u_{j,n+2} + 4u_{j,n+1} - 10u_{j,n} + 11u_{j,n-1} - \frac{7}{4}u_{j,n-2}}{-9\Delta t} \tag{3.96}$$

and it may be shown that this approximation has a truncation error of order $0(\Delta t^4)$. Substituting (3.95) and (3.96) into (3.93) we obtain the relation

$$\begin{aligned}
&\left(\frac{5}{6}\right)\left\{u_{j,n+3} - \left(\frac{17}{4}\right)u_{j,n+2} + 4u_{j,n+1} + 11u_{j,n-1} - \left(\frac{7}{4}\right)u_{j,n-2}\right\}\\
&+ \frac{\sigma(x,t)}{12\sigma(x+h,t)}\left\{u_{j+1,n+3} - \left(\frac{17}{4}\right)u_{j+1,n+2} + 4u_{j+1,n+1}\right.\\
&\left.+ 11u_{j+1,n-1} - \left(\frac{7}{4}\right)u_{j+1,n-2}\right\} + \frac{\sigma(x,t)}{12\sigma(x-h,t)}\left\{u_{j-1,n+3}\right.\\
&\left.- \left(\frac{17}{4}\right)u_{j-1,n+2} + 4u_{j-1,n+1} + 11u_{j-1,n-1} - \left(\frac{7}{4}\right)u_{j-1,n-2}\right\}\\
&+ \left\{9\alpha\sigma(x,t) - \left(\frac{5}{6}\right)\frac{\sigma(x,t)}{\sigma(x+h,t)}\right\}u_{j+1,n} - \left\{\frac{25}{3} + 18\alpha\sigma(x,t)\right\}u_{j,n}\\
&+ \left\{9\alpha\sigma(x,t) - \left(\frac{5}{6}\right)\frac{\sigma(x,t)}{\sigma(x-h,t)}\right\}u_{j-1,n} = 0,
\end{aligned} \tag{3.97}$$

where $h = \Delta x$ and $\alpha = \Delta t/h^2$.

Although this expression would appear at a first glance to be rather complicated, as we shall now show our problem may be reduced to one of finding the required solution of a fifth order linear recurrence relation and this may be done using certain of the techniques developed in previous sections. We first define a tri-diagonal matrix $\mathbf{R}(a, b, c)$ by the relation

$$\mathbf{R}(a,b,c) = \begin{bmatrix} b & c & & & & \\ a & b & c & & & \\ & & \cdot & \cdot & \cdot & \\ & & & \cdot & \cdot & \cdot \\ & & & a & b & c \\ & & & & a & b \end{bmatrix}$$

and a vector $\mathbf{u}_n$ by $\mathbf{u}_n = (u_{1,n}, u_{2,n}, \ldots, u_{1/(\Delta x)+1,n})^T$. We now rewrite the finite difference scheme (3.97) in the form

$$\mathbf{A}(j,n)u_{n+3} + \mathbf{B}(j,n)u_{n+2} + \mathbf{C}(j,n)u_{n+1} + \mathbf{D}(j,n)u_n + \mathbf{E}(j,n)u_{n-1} + \mathbf{F}(j,n)u_{n-2} = 0, \qquad (3.98)$$

where

$$\mathbf{A}(j,n) = \mathbf{R}\left(\frac{\sigma(x,t)}{12\sigma(x-h,t)}, \frac{5}{6}, \frac{\sigma(x,t)}{12\sigma(x+h,t)}\right),$$

$$\mathbf{B}(j,n) = -(\tfrac{17}{4})\mathbf{A}(j,n), \qquad \mathbf{C}(j,n) = 4\mathbf{A}(j,n),$$

$$\mathbf{E}(j,n) = 11\mathbf{A}(j,n), \qquad \mathbf{F}(j,n) = (-\tfrac{7}{4})\mathbf{A}(j,n),$$

and

$$\mathbf{D}(j,n) = \mathbf{R}\left(9\alpha\sigma(x,t) - \left(\frac{5}{6}\right)\frac{\sigma(x,t)}{\sigma(x-h,t)}, -\frac{25}{3} - 18\alpha\sigma(x,t), 9\alpha\sigma(x,t) - \left(\frac{5}{6}\right)\frac{\sigma(x,t)}{\sigma(x-h,t)}\right).$$

Thus we have reduced our problem to one of finding the required solution of the linear system of recurrence relations (3.98), where we note that all of the coefficients of this relation are tri-diagonal matrices, and we may generate this solution using one of the schemes developed in Chapter 2.

Two obvious candidates for the solution of this problem are:

(a) the direct boundary value approach developed in Section 2.5;
(b) the Gauss–Seidel iterative approach.

For demonstration purposes we consider (a) first of all and we illustrate this approach by setting $N = 4$. The equations which define the required vectors $\mathbf{u}_2$ and $\mathbf{u}_3$ are given by

$$\mathbf{C}(j,2)\mathbf{u}_3 + \mathbf{D}(j,2)\mathbf{u}_2 = -\mathbf{E}(j,2)\mathbf{u}_1 - \mathbf{F}(j,2)\mathbf{u}_0,$$

$$\mathbf{D}(j,3)\mathbf{u}_3 + \mathbf{E}(j,3)\mathbf{u}_2 = -\mathbf{F}(j,3)\mathbf{u}_1.$$

Eliminating $\mathbf{u}_3$ we have

$$[\mathbf{C}(j,2)\mathbf{E}(j,3) - \mathbf{D}(j,3)\mathbf{D}(j,2)]\mathbf{u}_2 = \mathbf{D}(j,3)[\mathbf{E}(j,2)\mathbf{u}_1 + \mathbf{F}(j,2)\mathbf{u}_0] - \mathbf{F}(j,3)\mathbf{C}(j,2)\mathbf{u}_1,$$

where the vectors $\mathbf{u}_0$ and $\mathbf{u}_1$ are assumed known. Thus in order to find $\mathbf{u}_2$ we need to generate the solution of a system of algebraic equations where the coefficient matrix is no longer tri-diagonal. As we allow the value of N to increase the coefficient matrices of the equations defining the $\mathbf{u}_i$ become less and less sparse and so, in general, this approach requires a considerable computational effort. If, however, we adopt approach (b) we may use the iteration scheme

$$\mathbf{A}(j,n)\mathbf{u}_{n+3}^{(p-1)} + \mathbf{B}(j,n)\mathbf{u}_{n+2}^{(p-1)} + \mathbf{C}(j,n)\mathbf{u}_{n+1}^{(p-1)} + \mathbf{D}(j,n)\mathbf{u}_{n}^{(p)} + \mathbf{E}(j,n)\mathbf{u}_{n-1}^{(p)} + \mathbf{F}(j,n)\mathbf{u}_{n-2}^{(p)} = 0. \quad (3.99)$$

An examination of this scheme shows that in order to solve for $\mathbf{u}_n^{(p)}$ we need to solve a tri-diagonal system of algebraic equations and this can be done very efficiently. In view of these considerations we will continue to use approach (b). In order to use iteration scheme (3.99), which behaves essentially like a three level integration scheme, we need two rows of solution values to be specified at consecutive time steps as our initial conditions. Normally only one row of initial conditions, $\mathbf{u}_0$ say, will be available and this forces us to use a separate "starting procedure" to generate an extra row of conditions. We shall return to this problem at a later stage and will assume for the time being that both $\mathbf{u}_0$ and $\mathbf{u}_1$ are available.

In order that we may be able to use scheme (3.99) in practice we also require approximations $\mathbf{u}_n^{(0)}$ to $\mathbf{u}_n$ to be available for all $n \geq 3$. Thus the complete set of initial conditions which we require is

$$\left.\begin{aligned} \mathbf{u}_0^{(p)} &= \mathbf{u}_0 \\ \mathbf{u}_1^{(p)} &= \mathbf{u}_1 \end{aligned}\right\} \quad \text{for all } p \geq 1,$$

$$\mathbf{u}_n^{(0)} = \mathbf{K}(n) \qquad n \geq 3.$$

Here $\mathbf{K}(n)$ is some sequence of initial approximations to the required solution $\mathbf{u}_n$ and clearly $\mathbf{K}(n)$ need not be completely specified initially but may be built up as the computation proceeds. We shall return to this aspect of our scheme at a later stage. In order to minimise the amount of storage space required, which is of course a very important practical consideration when dealing with partial differential equations, we calculate the sequence of iterates in the order defined by Fig. 3.7. It is clear that if the sequence $\{\mathbf{u}_n^{(p)}\}$ generated by scheme (3.99) converges in the sense that

$$\lim_{p\to\infty} \mathbf{u}_n^{(p)} = \hat{\mathbf{u}}_n \quad \text{for all} \quad n = 2, 3, \ldots$$

then the sequence $\hat{\mathbf{u}}_n$ will also satisfy (3.98). We now examine the stability properties of iterative scheme (3.99). Applying this scheme to the simple heat conduction equation

$$\frac{\partial u}{\partial t} = \sigma \frac{\partial^2 u}{\partial x^2}, \qquad \sigma = \text{constant} > 0$$

we obtain a relation of the form (3.98) where now the coefficients are all constant tri-diagonal matrices. Suppose now that we denote the required solution of (3.98) at the grid point $x = j\Delta x, t = n\Delta t$ by $u_{j,n}$. Due to the effect of

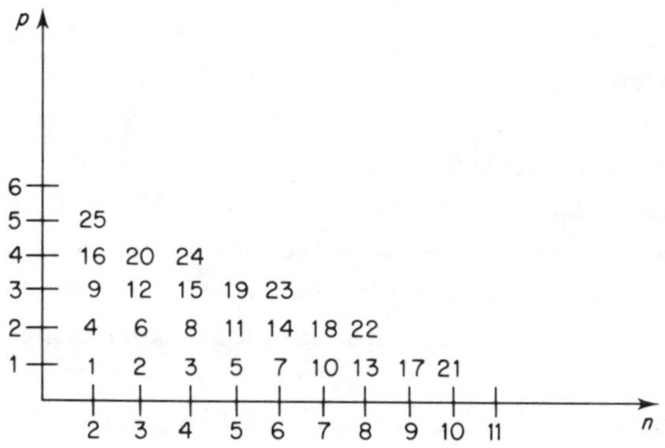

FIG. 3.7 Order of computation of iterates for (3.99).

rounding we actually obtain a solution $\tilde{u}_{j,n}$ where

$$u_{j,n} = \tilde{u}_{j,n} + \varepsilon_{j,n}$$

for some error sequence $\varepsilon_{j,n}$. Assuming that the given initial conditions are exact, that is $\varepsilon_{0,n} = \varepsilon_{1/\Delta x, n} = 0$, for all n, we may write the error sequence $\varepsilon_{j,n}$, which clearly satisfies (3.98), in the form

$$\varepsilon_{j,n} = \lambda^{n\Delta t} \sin(mj\Delta x), \quad m \equiv n\pi$$

where

$$(\tfrac{5}{6} + (\tfrac{1}{6})\cos(m\Delta x))(\lambda^5 - (\tfrac{17}{4})\lambda^4 + 4\lambda^3 + 11\lambda - \tfrac{7}{4})$$
$$- (\tfrac{25}{3} + 18\alpha\sigma - (18\alpha\sigma - \tfrac{5}{3})\cos(m\Delta x))\lambda^2 = 0. \qquad (3.100)$$

If we were to generate the required solution by direct forward recurrence in the usual way, the condition for the error $\varepsilon_{j,n}$ not to grow with n would be $|\lambda_i| \leq 1$, $i = 1, 2, 3, 4, 5$, where the λ_i's are the five roots, assumed distinct, of (3.100) and this is of course just the von Neumann necessary condition for stability. If, instead of generating the required solution using a direct method, we use scheme (3.99), the relation corresponding to (3.100) becomes

$$(\tfrac{5}{6} + \tfrac{1}{6}\cos(m\Delta x))([\lambda^{(p-1)}]^5 - \tfrac{17}{4}[\lambda^{(p-1)}]^4 + 4[\lambda^{(p-1)}]^3$$
$$+ 11\lambda^{(p)} - \tfrac{7}{4}) - (\tfrac{25}{3} + 18\alpha\sigma - (18\alpha\sigma - \tfrac{5}{3})\cos(m\Delta x))[\lambda^{(p)}]^2 = 0 \quad (3.101)$$

It is clear that the solutions of (3.100) satisfy the linear recurrence relation

$$(\tfrac{5}{6} + \tfrac{1}{6}\cos(m\Delta x))(y_{n+5} - (\tfrac{17}{4})y_{n+4} + 4y_{n+3} + 11y_{n+1} - (\tfrac{7}{4})y_n)$$
$$- (\tfrac{25}{3} + 18\alpha\sigma - (18\alpha\sigma - \tfrac{5}{3})\cos(m\Delta x))y_{n+2} = 0 \quad (3.102)$$

and our iteration scheme corresponds to the generation of a solution of this equation using the scheme

$$(\tfrac{5}{6} + \tfrac{1}{6}\cos(m\Delta x))(y_{n+5}^{(p-1)} - (\tfrac{17}{4})y_{n+4}^{(p-1)} + 4y_{n+3}^{(p-1)} + 11y_{n+1}^{(p)} - (\tfrac{7}{4})y_n^{(p)})$$
$$- (\tfrac{25}{3} + 18\alpha\sigma - (18\alpha\sigma - \tfrac{5}{3})\cos(m\Delta x))y_{n+2}^{(p)} = 0. \quad (3.103)$$

We shall define the interval of stability of scheme (3.103) to be that interval r of the positive real axis for which all solutions of (3.103) tend to zero as $n \to \infty$ for all $\alpha\sigma \in r$. It seems difficult to derive a useful algebraic expression defining the region of stability of scheme (3.103) so the procedure adopted to determine this stability region is essentially a numerical one. The procedure actually adopted in practice is to compute the solution of (3.103) for values of $\cos(m\Delta x)$ in the range $-1{\cdot}0(0{\cdot}1)1{\cdot}0$ and for increasing values of $\alpha\sigma$. It is clear that the asymptotic properties of the solutions of (3.103) are independent of the choice of initial conditions y_0, y_1 since, if there is a solution of (3.103) which ultimately increases, then a component of this solution will be brought into the solution which we actually calculate due to the effect of rounding no matter what initial conditions are used. The initial conditions actually used in the computation of the stability regions are $y_0 = 0$, $y_1 = 1$. The stability properties of (3.103) would seem at first sight to depend on the sequence of initial approximations $\mathbf{K}(n)$ but numerical experience so far gained in calculating these regions of absolute stability does seem to indicate that for most bounded sequences $\mathbf{K}(n)$ scheme (3.103) is unconditionally stable. Quite wide experience gained on a variety of practical problems also seems to indicate the possibility of our scheme being unconditionally stable. A simple necessary condition for the stability of our scheme may be derived by splitting up the solution of (3.103) into a particular solution and a set of two complementary functions. For any positive integer p the complementary functions of (3.103) satisfy the quadratic

$$\left(\tfrac{25}{3} + 18\alpha\sigma - (18\alpha\sigma - \tfrac{5}{3})\cos(m\Delta x)\right)y^2 - 11\left(\tfrac{5}{6} + \tfrac{1}{6}\cos(m\Delta x)\right)y$$
$$+ \tfrac{7}{4}\left(\tfrac{5}{6} + \tfrac{1}{6}\cos(m\Delta x)\right) = 0. \qquad (3.104)$$

Applying Schur's theorem (Lambert, 1973, p. 78) it follows that a necessary and sufficient condition for the roots of this quadratic to be less than one in modulus, which is clearly a necessary condition for the stability of our overall scheme, is

$$\tfrac{7}{4}\left(\tfrac{5}{6} + \tfrac{1}{6}\cos(m\Delta x)\right) \leq \tfrac{25}{3} + 18\alpha\sigma - (18\alpha\sigma - \tfrac{5}{3})\cos(m\Delta x)$$

and

$$11\left(\tfrac{5}{6} + \tfrac{1}{6}\cos(m\Delta x)\right) \leq \tfrac{25}{3} + 18\alpha\sigma - (18\alpha\sigma - \tfrac{5}{3})\cos(m\Delta x)$$
$$+ \tfrac{7}{4}\left(\tfrac{5}{6} + \tfrac{1}{6}\cos(m\Delta x)\right). \qquad (3.105)$$

It is trivial to verify algebraically that these conditions are satisfied for all α. In the one-dimensional case we find that the performance of our schemes is greatly improved if we use an accurate predictor to generate the values $\{y_n^{(0)}\}$. This conclusion also seems to be valid in the two-dimensional case and we shall comment on this later. Before presenting any numerical results we consider a somewhat different approach to the solution of our problem. We make the same approximation to the space derivative as was given by (3.95) and, substituting this into (3.93), we obtain the system of first order ordinary differential equations

$$\frac{du}{dt}(x,t) = \frac{u(x+h,t) - 2u(x,t) + u(x-h,t)}{(h)^2}$$
$$-\frac{1}{12}\left\{\frac{1}{\sigma(x+h,t)}\frac{du}{dt}(x+h,t)\right.$$
$$\left.-\frac{2}{\sigma(x,t)}\frac{du}{dt}(x,t) + \frac{1}{\sigma(x-h,t)}\frac{du}{dt}(x-h,t)\right\}.$$

Defining a positive integer n which is such that $(n+1)h = 1$ and putting $u(t) = [u(h,t), u(2h,t), \ldots, u(nh,t)]^T$ this system of ordinary differential equations may be written in the matrix form

$$\mathbf{A}\frac{d\mathbf{u}}{dt}(t) = \frac{1}{h^2}\hat{\mathbf{B}}\mathbf{u}(t) + \hat{\mathbf{C}},$$

where

$$\mathbf{A} = \begin{bmatrix} 1-\dfrac{1}{6\sigma(h,t)} & \dfrac{1}{12\sigma(2h,t)} & & & \mathbf{0} \\ \dfrac{1}{12\sigma(h,t)} & 1-\dfrac{1}{6\sigma(2h,t)} & \dfrac{1}{12\sigma(3h,t)} & & \\ & \cdot & \cdot & \cdot & \\ \mathbf{0} & & \cdot & \cdot & \cdot \end{bmatrix}$$

$$\hat{\mathbf{B}} = \begin{bmatrix} -2 & 1 & & & & 0 \\ 1 & -2 & 1 & & & \\ & \cdot & \cdot & \cdot & & \\ & & \cdot & \cdot & \cdot & \\ & & & 1 & -2 & 1 \\ 0 & & & & 1 & -2 \end{bmatrix}$$

$$\hat{\mathbf{C}} = \begin{bmatrix} \dfrac{u(0,t)}{h^2} - \dfrac{1}{12\sigma(0,t)}\dfrac{\partial u}{\partial t}(0,t) \\ 0 \\ \vdots \\ 0 \\ \dfrac{u(1,t)}{h^2} - \dfrac{1}{12\sigma(1,t)}\dfrac{\partial u}{\partial t}(1,t) \end{bmatrix}$$

In order to evaluate the two time derivatives $\dfrac{\partial u}{\partial t}(0,t)$ and $\dfrac{\partial u}{\partial t}(1,t)$ it is necessary to use a suitable finite difference approximation. The one which we actually use in practice is given by

$$\frac{\partial u(0,t)}{\partial t} = \frac{u(0,t+3\Delta t) - 6u(0,t+2\Delta t) + 18u(0,t+\Delta t) - 10u(0,t) - 3u(0,t-\Delta t)}{12\Delta t}$$

with a similar one for $\dfrac{\partial u}{\partial t}(1,t)$. In this way we may reduce our original problem to a system of first order ordinary differential equations of the form

$$\mathbf{A}\frac{\mathrm{d}\mathbf{u}}{\mathrm{d}t} = \mathbf{B}\mathbf{u} + \mathbf{C}. \tag{3.106}$$

In order to determine our final solution it is now necessary to integrate (3.106). Since the truncation error of the space discretisation is $0(\Delta x^4)$ we shall integrate (3.106) using a fourth order method so that the finite difference approximation is fourth order in both the space and time variables. The scheme which we actually use for this purpose is the block cyclic scheme (see Cash, 1977c) which integrates forward in blocks of four equal time steps. Since this method has already been fully described in Cash (1977c) and since there are no additional problems present in the integration of (3.106) which were not present previously we shall not describe our algorithm any further.

Finally while considering semi-discretisation schemes of this type we draw attention to the sliding difference algorithm of Loeb and Schiesser (1974). In their approach, space derivatives away from the boundary are approximated by suitable central difference representations while space derivatives near to the boundary are approximated by either forward or backward differences, whichever is appropriate, so that only internal points are used. This approach does not require the time derivative to be approximated on the boundary and this allows us to extend our approach to problems with derivative boundary conditions. Thus an obvious extension of our method would be to use the Loeb–Schiesser fourth order sliding difference approximation to approximate the space derivative and then to integrate the resulting system of ordinary differential equations using our fourth order block integration method. Again this is a straightforward extension of previous analysis and so will not be considered further. It is however worth commenting that this overall approach has proved to be useful for the integration of parabolic partial differential equations.

As a preliminary to presenting some numerical results we mention briefly one of the important practical aspects associated with comparing algorithm (3.99) with more conventional algorithms. The important point to note is that when using scheme (3.99) we need to compute an extra row of initial conditions at the time step $t = \Delta t$, but as we shall now show this problem is not nearly as serious as it may seem at first sight. Suppose for example that we wish to compare the performance of our iterative algorithm with that of a standard one-step finite difference scheme S_1. What we would in fact do is to use scheme S_1 to calculate the extra row of initial conditions needed by our iterative algorithm (so that the two algorithms are identical at this stage) and then we would compare the two algorithms for $t > \Delta t$. Thus, if we believe that a one-step scheme $\hat{s}$ is in some vague sense "the best" standard finite difference scheme, our composite integration procedure would consist of using $\hat{s}'$ to compute the solution at time Δt and then scheme (3.99) would be used to compute the solution for all $t > \Delta t$. In view of this, starting does not present nearly as severe a problem as we may at first think.

We now conclude this chapter with a numerical example illustrating the use of iteration scheme (3.99). The particular problem we consider is the integration of the linear parabolic equation

$$\frac{\partial u}{\partial t} = (1 + t)\frac{\partial^2 u}{\partial x^2} \tag{3.107}$$

with initial conditions

$$u(x, 0) = \sum_{n=1}^{m} \exp(-\tfrac{1}{2}n^2)\cos nx \qquad 0 \leq x \leq 1,$$

$$\left.\begin{aligned} u(0, t) &= \sum_{n=1}^{m} \exp(-\tfrac{1}{2}n^2(1 + t)^2), \\ u(1, t) &= \sum_{n=1}^{m} \cos(n)\exp(-\tfrac{1}{2}n^2(1 + t)^2), \end{aligned}\right\} \quad t \geq 0$$

where the two values of m considered are $m = 1, 5$. The true solution of this problem is given by

$$u(x, t) = \sum_{n=1}^{m} \exp(-\tfrac{1}{2}n^2(1 + t)^2)\cos(nx).$$

The two schemes for the solution of this problem which we compare are:

(a) the Crank–Nicolson scheme with $\Delta x = \Delta t$;

(b) scheme (3.99) with $\Delta x = \Delta t$.

Numerical solutions of (3.107) are computed using (a) and (b) for a range of values of Δx and in each case the relative accuracy, p, in the L_2 norm of the solution is computed at $t = 1{\cdot}0$ using the known analytic solution. For demonstration purposes we shall always assume that the analytic solution is known at time $t = \Delta t$ and we compare our solution with that obtained using Crank–Nicolson for times $t > \Delta t$. As mentioned previously there is no loss in generality in adopting this approach since when carrying out a genuine comparison we first use the Crank–Nicolson scheme to compute the solution at time $t = \Delta t$ before switching to iteration scheme (3.99) to compute solutions for time $t > \Delta t$. When using scheme (3.99) the initial sequence of approximations $\mathbf{K}(n) \equiv 0$ is used and at each step iterates are computed until the L_2 norm of two successive iterates differs by less than 10^{-7}. In order to keep the storage space required to a minimum successive iterates are computed in the order defined by Fig. 3.7. In Fig. 3.8 we give the graph of the

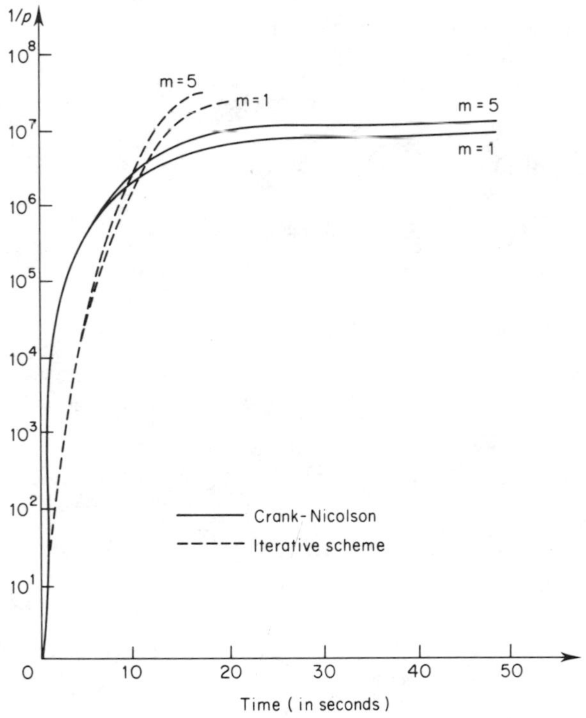

FIG. 3.8 Times to achieve a relative accuracy of p for the integration of (3.107).

time taken to achieve a relative accuracy of p in the L_2 norm of the solution for a range of values of p. As can be seen the Crank–Nicolson method is the more efficient of the two for lower accuracy requirements but, as the required degree of accuracy increases, the Crank–Nicolson scheme rapidly becomes inefficient compared with scheme (3.99). In cases where relatively low orders of accuracy are required the performance of scheme (3.99) may be considerably improved by

(a) using an explicit predictor or polynomial extrapolation based on a polynomial of degree 1 or 2 to generate $\mathbf{K}(n)$ (cf. the one-dimensional case described in Section 3.3);

(b) relaxing the restriction that successive iterates at a particular time level should differ by 10^{-7} to the requirement that they should differ by p say.

Although the performance of scheme (3.99) is often considerably improved when these two modifications are made (computation was more than halved in some cases for low orders of accuracy) we shall not pursue this approach any further since our main aim is merely to establish the progressively increasing efficiency of (3.99) compared with Crank–Nicolson for the particular test problem (3.107) as increasing degrees of precision are required. In Table 3.11 we list the actual times taken by both schemes to achieve some of the higher degrees of accuracy. It is interesting to note that if, for example, a relative accuracy of 10^{-7} is required for the case $m = 5$ the Crank–Nicolson scheme requires 500 grid points for each value of t whereas scheme (3.99) requires only 25. This is clearly an important consideration if these schemes are extended to be applicable to problems in more than one space dimension.

TABLE 3.11

		Crank–Nicolson		
	$m = 1$		$m = 5$	
$1/\Delta x$	Accuracy	Time taken (sec)	Accuracy	Time taken (sec)
50	$0{\cdot}833 \times 10^{-5}$	1·247	$0{\cdot}103 \times 10^{-4}$	1·530
100	$0{\cdot}219 \times 10^{-5}$	2·536	$0{\cdot}237 \times 10^{-5}$	3·017
150	$0{\cdot}970 \times 10^{-6}$	4·695	$0{\cdot}131 \times 10^{-5}$	5·818
225	$0{\cdot}430 \times 10^{-6}$	9·573	$0{\cdot}510 \times 10^{-6}$	10·471
300	$0{\cdot}242 \times 10^{-6}$	16·261	$0{\cdot}282 \times 10^{-6}$	17·492
500	$0{\cdot}870 \times 10^{-7}$	43·747	$0{\cdot}102 \times 10^{-6}$	45·526
		Scheme (3.99)		
10	$0{\cdot}122 \times 10^{-4}$	3·499	$0{\cdot}312 \times 10^{-5}$	3·563
15	$0{\cdot}236 \times 10^{-5}$	4·929	$0{\cdot}693 \times 10^{-6}$	4·931
20	$0{\cdot}740 \times 10^{-6}$	6·834	$0{\cdot}234 \times 10^{-6}$	6·935
25	$0{\cdot}295 \times 10^{-6}$	8·729	$0{\cdot}106 \times 10^{-6}$	8·860
30	$0{\cdot}128 \times 10^{-6}$	11·260	$0{\cdot}670 \times 10^{-7}$	11·137

4

Some Iterative Algorithms Without Truncation Error and their Connection with Singular Perturbation Methods for the Solution of O.D.E.s

In the previous two chapters we were able to show how precisely defined iteration techniques may be used to produce certain non-dominant solutions of linear recurrence relations for which direct methods of solution are unstable except over a very small interval. The main function of these iterative techniques is to repose the original *initial value problem* as a *boundary value problem*. These two problems are equivalent in the sense that the boundary value problem has a solution which closely approximates the solution of the original initial value problem over some prescribed range of the independent variable. If, for example, we consider the linear recurrence relation

$$y_{n+1} + py_n + qy_{n-1} = 0, \qquad y_0 = \bar{y},$$

where for the sake of simplicity we assume that both p and q are constants, we may in certain circumstances generate a non-dominant solution of this equation using the iteration scheme

$$y_{n+1}^{(t-1)} + py_n^{(t)} + qy_{n-1}^{(t)} = 0, \qquad t = 1, 2, \ldots,$$

$$y_n^{(0)} \equiv 0 \text{ for all } n > 0, \qquad y_0^{(t)} \equiv \bar{y} \text{ for all } t \geq 0.$$

This iteration scheme corresponds to the solution of an appropriate tridiagonal system of linear algebraic equations using a Gauss–Seidel iteration scheme. Another way of describing the role of this iteration scheme is to say that it reduces the original homogeneous *second order* equation to an inhomogeneous one of the *first order* the only solution of which is the required non-dominant solution of the original equation. The main purpose of the present chapter is to examine how this approach may be extended to deal with ordinary differential equations so that the iteration scheme is applied directly to the differential equation itself and not to an approximating discretisation scheme. As a simple example we consider the linear, second order ordinary differential equation

$$y'' + py' + qy = 0, \qquad \text{where } ' \equiv \frac{\mathrm{d}}{\mathrm{d}x}.$$

Following the procedure just described for linear recurrence relations we could seek to reduce this differential equation to one of the first order by using the iteration

$$\frac{\mathrm{d}^2 y^{(t-1)}}{\mathrm{d}x^2} + p\frac{\mathrm{d}y^{(t)}}{\mathrm{d}x} + qy^{(t)} = 0. \tag{4.1}$$

Given a sequence of initial approximations $\mathrm{d}^2y^{(0)}/\mathrm{d}x^2$ for all x in the range of integration we may substitute the value $t = 1$ into (4.1) to obtain an expression for $\mathrm{d}y^{(1)}/\mathrm{d}x$. Differentiating this relation with respect to x and substituting in (4.1) for $t = 2$ we obtain an expression for $\mathrm{d}y^{(2)}/\mathrm{d}x$ and this procedure may be continued for $t = 3, 4, \ldots$ until, in "favourable" circumstances, convergence in some appropriate sense is achieved. This class of techniques has a much more limited range of application than those developed in the previous chapter for the numerical solution of ordinary differential equations using finite difference schemes. The main reason for this is that, when dealing with discretisation schemes, we are usually able to choose the particular schemes which we use to take a form which is such that the iterative algorithms which we use to generate the required solution of the discretisation scheme are rapidly convergent. In contrast with this we find that schemes of the form (4.1) are only of use if the original differential equation has a special structure. Nevertheless, iterative algorithms of the form (4.1) do have important practical applications in the numerical solution of certain classes of ordinary differential equations of practical importance and in this chapter we shall investigate these algorithms and examine their relation to certain other classes of integration procedures currently in use.

4.1 Iterative solution of linear second order O.D.E.s

In order to fix our main ideas we consider first the scalar, linear, second order ordinary differential equation

$$\frac{d^2y}{dx^2} + p(x)\frac{dy}{dx} + q(x)y = 0, \tag{4.2}$$

where $p(x)$ and $q(x)$, and their first derivatives with respect to x, are well defined for all x in the range of interest. We further assume that $p(x) \neq 0$ for any x in the range of integration. We shall not impose any initial or boundary conditions for the time being but we will find out later which conditions are the most convenient ones to impose in order that the required solution may be specified uniquely. We shall assume for the time being that $y(x)$ does not have any zeros in the range of interest so that we may rewrite (4.2) in the more convenient form

$$y''/y + p(x)y'/y + q(x) = 0. \tag{4.3}$$

If zeros of y do occur at a finite sequence of points $\{x_i\}$, $i = 1, 2, \ldots, N$, we may apply our iteration scheme in a range which stops just short of x_1, we may then integrate through the point x_1 using a standard integration procedure, and we may then apply our iteration scheme in a range $(x_1 + \eta, x_2 - \eta)$ for some sufficiently small η. This procedure may be continued in a straightforward manner and in this way all points at which $y(x) = 0$, and their neighbourhoods, may be eliminated from the region of application of (4.3). Before considering an iteration scheme for the solution of (4.3) we first define the precise solution of (4.2) which we require. The general solution, $y(x)$, of (4.2) may be written in the form

$$y(x) = c_1 y_1(x) + c_2 y_2(x), \tag{4.4}$$

where $y_1(x)$ and $y_2(x)$ are any two solutions of (4.2) which are such that the vectors $[y_1(x_0), y_1'(x_0)]^T$, $[y_2(x_0), y_2'(x_0)]^T$ are linearly independent for some x_0 in the range of integration and c_1, c_2 are arbitrary constants. We assume that it is possible to choose the basis solutions $y_1(x)$ and $y_2(x)$ such that

$$\left|\frac{y_1''(x)}{y_1(x)}\right| \gg \left|\frac{y_2''(x)}{y_2(x)}\right| \tag{4.5}$$

for all x in the range of interest and we shall assume from now on that we wish to compute a solution of (4.2) which is directly proportional to $y_2(x)$.

Following very closely the procedure adopted for linear recurrence relations we attempt to generate a solution of (4.3) using the iteration scheme

$$(y''/y)^{(t-1)} + p(x)(y'/y)^{(t)} + q(x) = 0, \qquad t = 1, 2, \ldots, \tag{4.6}$$

where here $y \equiv y(x)$. What we are in effect trying to do is to eliminate the unwanted solution, $y_1(x)$, of (4.3) from the solution which we actually obtain by reducing (4.3) to a first order ordinary differential equation the only solution of which is $y_2(x)$. In order to be able to use scheme (4.6) in practical applications we first need to make an initial approximation to $(y''/y)^{(0)}$ at all points in the region of integration. If we constrain this approximation to be very small initially we may hope to obtain convergence to a solution which is directly proportional to $y_2(x)$ by virtue of relation (4.5); that is, we choose the quotient $(y''/y)^{(0)}$ to be everywhere small initially with the aim of finding a solution of (4.2) for which $|y''(x)/y(x)|$ is relatively small throughout the range of integration no matter what initial conditions are imposed on $y(x)$. This demand that $|y''/y|$ should remain relatively small everywhere in the range of integration serves as one of our (rather imprecise) boundary conditions. This boundary condition highlights the rather limited range of applicability of our algorithm since it is clear that we can only expect to solve those problems for which a condition of the form just described picks out precisely which basis solution $y_2(x)$ we require. Suppose now that for demonstration purposes we use the sequence of initial approximations $(y''/y)^{(0)} \equiv 0$. Substituting this sequence into (4.6) for $t = 1$ we obtain

$$(y'/y)^{(1)} = -q(x)/p(x) = A_1(x) \quad \text{say.} \tag{4.7}$$

Differentiating relation (4.7) formally with respect to x we obtain

$$(y''/y)^{(1)} - [(y'/y)^{(1)}]^2 = A_1'(x),$$

where $A_1'(x) \equiv \partial A_1(x)/\partial x$ and so $(y''/y)^{(1)} = \{A_1(x)\}^2 + A_1'(x)$.

Substituting this expression for $(y''/y)^{(1)}$ into (4.6) for $t = 2$ we obtain

$$\begin{aligned}(y'/y)^{(2)} &= -\frac{q(x)}{p(x)} - \left\{\frac{A_1'(x) + \{A_1(x)\}^2}{p(x)}\right\},\\ &= A_2(x) \quad \text{say.}\end{aligned}$$

Continuing this process for increasing integer values of the iteration parameter

t we obtain a sequence of approximations of the form

$$\left.\begin{aligned} &(y'/y)^{(t)} = A_t(x), \\ \text{where}\quad &A_t(x) = -\frac{q(x)}{p(x)} - \left\{\frac{A'_{t-1}(x) + \{A_{t-1}(x)\}^2}{p(x)}\right\}, \qquad t = 1, 2, \ldots, \\ \text{with}\quad &A_0(x) = A'_0(x) \equiv 0. \end{aligned}\right\} \tag{4.8}$$

This iteration scheme may be continued until either a maximum allowable number of iterates have been computed (and by implication convergence has not occurred) or until some integer i is reached for which

$$\left|\frac{A_i(x) - A_{i-1}(x)}{A_{i-1}(x)}\right| < \varepsilon, \text{ if relative accuracy is required,}$$

or

$$|A_i(x) - A_{i-1}(x)| < \varepsilon, \text{ if absolute accuracy is required,}$$

for all points x in the range of integration. Here ε is some prescribed local accuracy tolerance. Assuming that the iteration scheme does converge to the prescribed degree of precision, the final sequence of iterates $y' = A_i(x)y$ may be integrated using a standard integration procedure to give the final solution of our problem. In order to uniquely define the required solution of this first order differential equation we need to specify one initial condition of the form $y(x_0) = y_0$. Thus, summing up, the boundary conditions which we use for the numerical integration of (4.2) are

$$\left|\frac{y''(x)}{y(x)}\right| \ll \left|\frac{y''_1(x)}{y_1(x)}\right| \text{ for any initial conditions imposed on } y(x),$$

$$y(x_0) = y_0,$$

where $y_1(x)$ is the unwanted solution of (4.2) having the property specified by (4.5) for all x. The first of these two boundary conditions ensures that the required solution is directly proportional to $y_2(x)$ and the second boundary condition tells us which multiple of $y_2(x)$ is required. These boundary conditions are somewhat imprecise and as such may not always specify the required solution uniquely especially if basis solutions cannot be chosen so that relation (4.5) holds. A particular case where it can be seen immediately that the given boundary conditions do specify the required solution uniquely is when the two basis solutions $y_1(x)$ and $y_2(x)$ may be chosen so that they are approximately exponentials of the form $e^{p_1 x}$ and $e^{p_2 x}$ with $|p_1| \gg |p_2|$. It is clear that in this case the solution specified by the boundary conditions will be directly proportional to $y_2(x)$.

As we shall now show, the convergence of scheme (4.6) is rather more difficult to analyse than the convergence of similar processes used for the numerical solution of linear recurrence relations in Chapter 2. This is due mainly to the fact that the intermediate solutions $y^{(t)}(x)$ do not satisfy the original differential equation and, as we shall show, it is only the finally converged sequence $y^{(i)}(x)$ which possesses this property. In order to understand the convergence properties of scheme (4.6) more fully we derive a relationship between successive errors. We first define an error $\eta_{t,n}$ at any point x_n as

$$\eta_{t,n} = \frac{y'(x_n)}{y(x_n)} - A_t(x_n), \tag{4.9}$$

where $y(x_n)$ is the analytic solution of (4.2) specified by the given boundary conditions. Since $\eta_{t,n}$ is a continuous function of x we may differentiate (4.9) formally with respect to this variable to obtain

$$\eta'_{t,n} = \frac{y''(x_n)}{y(x_n)} - \left(\frac{y'(x_n)}{y(x_n)}\right)^2 - A'_t(x_n). \tag{4.10a}$$

Using relation (4.9) we now have

$$\eta_{t,n} = \frac{y'(x_n)}{y(x_n)} - A_t(x_n),$$

$$= \frac{y'(x_n)}{y(x_n)} + \frac{q(x_n)}{p(x_n)} + \left\{\frac{A'_{t-1}(x_n) + \{A_{t-1}(x_n)\}^2}{p(x_n)}\right\} \qquad \text{from (4.8)},$$

$$= \frac{y'(x_n)}{y(x_n)} + \frac{q(x_n)}{p(x_n)} + \frac{1}{p(x_n)}\left\{\frac{y''(x_n)}{y(x_n)} - \left(\frac{y'(x_n)}{y(x_n)}\right)^2 - \eta'_{t-1,n}\right\}$$

$$+ \frac{1}{p(x_n)}\left\{\left(\frac{y'(x_n)}{y(x_n)}\right)^2 - 2\eta_{t-1,n}\frac{y'(x_n)}{y(x_n)} + \eta^2_{t-1,n}\right\} \qquad \text{from (4.9), (4.10a)},$$

$$= -\frac{1}{p(x_n)}\left\{\eta'_{t-1,n} + 2\frac{y'(x_n)}{y(x_n)}\eta_{t-1,n} - \eta^2_{t-1,n}\right\} \qquad \text{from (4.2)},$$

$$= -\frac{1}{p(x_n)}\left\{\eta'_{t-1,n} + 2A_{t-1}(x_n)\eta_{t-1,n} + \eta^2_{t-1,n}\right\} \qquad \text{from (4.9)}.$$

This expression shows the complicated relationship that exists between successive errors and highlights the difficulty in obtaining a useful convergence criterion. It is, however, of interest to note that one of these error terms may be removed in the following way.

Considering the above error relationship for $t = 2$ we have

$$\begin{aligned}\eta_{2,n} &= -\frac{1}{p(x_n)}\{\eta'_{1,n} + 2A_1(x_n)\eta_{1,n} + \eta^2_{1,n}\},\\ &= -\frac{1}{p(x_n)}\{\eta'_{1,n} + 2A_1(x_n)[\eta_{1,n} - \eta_{2,n} + \eta_{2,n}] + \eta^2_{1,n}\},\\ &= -\frac{1}{p(x_n)}\{\eta'_{1,n} + 2A_1(x_n)\eta_{2,n} + \eta^2_{1,n}\}\\ &\quad - \frac{2A_1(x_n)}{p(x_n)}(A_2(x_n) - A_1(x_n)),\end{aligned}$$

and so

$$\eta_{2,n} = -\frac{\{\eta'_{1,n} + \eta^2_{1,n}\}}{p(x_n) + 2A_1(x_n)} - \frac{2A_1(x_n)(A_2(x_n) - A_1(x_n))}{p(x_n) + 2A_1(x_n)}. \tag{4.10b}$$

Suppose now we define an error $\hat{\eta}_{2,n}$ as

$$\hat{\eta}_{2,n} = \frac{y'(x_n)}{y(x_n)} - A_2(x_n) + \frac{2A_1(x_n)[A_2(x_n) - A_1(x_n)]}{p(x_n) + 2A_1(x_n)}.$$

Then

$$\begin{aligned}\hat{\eta}_{2,n} &= \eta_{2,n} + \frac{2A_1(x_n)[A_2(x_n) - A_1(x_n)]}{p(x_n) + 2A_1(x_n)} && \text{from (4.9)},\\ &= -\left\{\frac{\eta'_{1,n} + \eta^2_{1,n}}{p(x_n) + 2A_1(x_n)}\right\} && \text{from (4.10b)}\end{aligned}$$

Thus if we define a new quantity $\hat{A}_2(x_n)$ as

$$\begin{aligned}\hat{A}_2(x_n) &= A_2(x_n) - \frac{2A_1(x_n)[A_2(x_n) - A_1(x_n)]}{p(x_n) + 2A_1(x_n)},\\ &= \frac{p(x_n)A_2(x_n) + 2\{A_1(x_n)\}^2}{p(x_n) + 2A_1(x_n)},\end{aligned}$$

we have

$$\hat{\eta}_{2,n} \equiv \frac{y'(x_n)}{y(x_n)} - \hat{A}_2(x_n) = -\left\{\frac{\eta'_{1,n} + \eta^2_{1,n}}{p(x_n) + 2A_1(x_n)}\right\}.$$

In general if we define these quantities recursively by

$$\left.\begin{aligned} A_t(x_n) &= -\frac{q(x_n)}{p(x_n)} - \left\{\frac{\hat{A}'_{t-1}(x_n) + \{\hat{A}_{t-1}(x_n)\}^2}{p(x_n)}\right\}, \\ &\hat{A}_0(x_n) = \hat{A}'_0(x_n) = 0, \\ \hat{A}_t(x_n) &= \frac{A_t(x_n)p(x_n) + 2\{\hat{A}_{t-1}(x_n)\}^2}{p(x_n) + 2\hat{A}_{t-1}(x_n)}, \\ \text{then} \qquad & \\ \hat{\eta}_{t,n} &= -\left\{\frac{\hat{\eta}'_{t-1,n} + \hat{\eta}^2_{t-1,n}}{p(x_n) + 2\hat{A}_{t-1}(x_n)}\right\}. \end{aligned}\right\} \tag{4.11}$$

It usually turns out that this iteration scheme converges faster than the one based on (4.6) and we shall provide some numerical evidence of this a little later. Although we are not able to derive a useful sufficient condition for convergence of either iteration scheme, we are able to say that if convergence does occur then it does so to a solution of the original problem. To see this we suppose that the sequence $A_t(x_n)$ considered as a function of t converges to the limit sequence $A(x_n)$. Then from (4.8) we have

$$A(x_n) = -\frac{q(x_n)}{p(x_n)} - \frac{1}{p(x_n)}\{A'(x_n) + \{A(x_n)\}^2\}$$

and this is precisely the condition which needs to be satisfied by $A(x_n)$ in order that the factor $y' = A(x_n)y$ should be a solution of our original differential equation (4.2). Although the theoretical analysis of iteration scheme (4.6) has proved to be very difficult, it turns out that in the constant coefficient case things are sufficiently simple for us to be able to obtain some concrete results. If we consider the constant coefficient equation

$$y'' + py' + qy = 0,$$

then the sequence A_t defined by (4.8) is

$$A_1 = -\frac{q}{p}$$

$$A_2 = A_1 - \frac{A_1^2}{p} = A_1\left(1 - \frac{A_1}{p}\right)$$

$$A_3 = A_1 - \frac{A_1^2}{p}\left(1 - \frac{A_1}{p}\right)^2 = A_1\left(1 - \frac{A_1}{p}\left(1 - \frac{A_1}{p}\right)^2\right)$$

$$A_4 = A_1 - \frac{A_1^2}{p}\left(1 - \frac{A_1}{p}\left(1 - \frac{A_1}{p}\right)^2\right)^2$$

$$= A_1\left(1 - \frac{A_1}{p}\left(1 - \frac{A_1}{p}\left(1 - \frac{A_1}{p}\right)^2\right)^2\right)$$

. . . .

. . . .

These expressions may now be regarded as functions of the single variable A_1/p only since each iterate may be written in the form $A_i/A_1 = f_i(A_1/p)$. By computing the limit of the sequence A_i/A_1 numerically for increasing i and for a range of values of the quantity A_1/p, which we have assumed to be real, it is found that this scheme converges in the range

$$-\tfrac{1}{4} \leq \frac{A_1}{p} < \tfrac{3}{4}.$$

The left hand inequality gives

$$-\tfrac{1}{4} \leq -\frac{q}{p^2},$$

i.e.

$$p^2 \geq 4q, \tag{1}$$

which is the condition for the roots of the polynomial $x^2 + px + q = 0$ to be real. The right hand inequality gives

$$3p^2 + 4q > 0, \tag{2}$$

and conditions (1), (2) give the inequalities to be satisfied by the coefficients of our differential equation (4.2) in order that our iteration scheme (4.6) should be convergent.

4.2 Extension of Olver's approach to the solution of O.D.E.s

In the previous section our main aim was to extend some of the algorithms for the numerical solution of linear recurrence relations which were derived in

Chapter 2 so that they may be applied directly to the numerical solution of linear, second order ordinary differential equations. Following this general approach we may develop an alternative numerical method for the solution of (4.2) which has a marked similarity with the Olver/Miller type of approach. This approach was described in Chapter 2 in connection with the solution of linear recurrence relations. If we formally differentiate equation (4.2) with respect to x we obtain

$$y''' + p(x)y'' + [p'(x) + q(x)]y' + q'(x)y = 0$$

which is of the general form

$$y''' + c_{2,1}(x)y'' + c_{2,2}(x)y' + c_{2,3}(x)y = 0.$$

Continuing this successive differentiation we obtain the sequence of relations

$$y^m + \sum_{i=0}^{m-1} c_{m-1,i+1}(x)y^{m-i-1} = 0 \tag{4.12a}$$

for all $m \geq 2$ where $y^j \equiv (d^jy)/(dx^j)$ with $y^0 \equiv y$. Suppose now that for some large integer value of m, $m = N$ say, we set $y^m = 0$. We are then left with the $N - 1$ linear equations obtained from (4.12a) by putting $m = 2, 3, \ldots, N$ in the N unknowns $y, y', y'', \ldots, y^{N-1}$. Writing these equations in matrix form we have

$$\begin{bmatrix} c_{1,1}(x) & 1 & & & \\ c_{2,2}(x) & c_{2,1}(x) & 1 & & \\ c_{3,3}(x) & c_{3,2}(x) & c_{3,1}(x) & 1 & \\ \cdot & & & & \\ \cdot & & & & \\ \cdot & & & & \\ c_{N-1,N-1}(x) & c_{N-1,N-2}(x) & \cdots\cdots & c_{N-1,2}(x) & c_{N-1,1}(x) \end{bmatrix} \begin{bmatrix} y' \\ y'' \\ y''' \\ \cdot \\ \cdot \\ \cdot \\ y^{N-1} \end{bmatrix}$$

$$= \begin{bmatrix} -c_{1,2}(x) \\ -c_{2,3}(x) \\ -c_{3,4}(x) \\ \cdot \\ \cdot \\ \cdot \\ -c_{N-1,N}(x) \end{bmatrix} y. \tag{4.12b}$$

If, starting from the last equation and working backwards, we successively eliminate $y^{N-1}, y^{N-2}, \ldots, y''$ we are finally left with a single, linear, first order ordinary differential equation of the form

$$y' = B^{[N]}(x)y.$$

Following the approach adopted by Miller (1952) in formulating his backward recurrence algorithm, we compute the sequences $B^{[N]}(x_n)$ for increasing values of the integer N until two successive sequences $B^{[N]}(x)$ and $B^{[N+1]}(x)$ are sufficiently close in some appropriate sense. When this is the case we integrate the equation $y' = B^{[N+1]}(x)y$ using any standard integration procedure to generate the final solution sequence $\{y_n\}$. This algorithm involves several important computational aspects which require a thorough investigation and we shall return to this later on in this section. As with the algorithm based on (4.6) which was described earlier in this chapter it seems difficult to derive a useful sufficient condition for the convergence of this Miller-type algorithm. Furthermore, in common with most other backward recurrence algorithms, we are not able to state positively that if convergence does occur in the sense that two successive sequences $B^{[j]}(x)$ differ by less than a prescribed amount then we have convergence to the required solution. (The convergence of backward recurrence algorithms as applied to the solution of difference equations has, however, recently been investigated by Zahar (1977) and the interested reader is referred to this paper and also to the references to earlier work contained in Zahar's paper.) Despite these problems this algorithm does have some useful practical applications and we shall return to it later in the section.

We now mention the possibility of extending Olver's approach, whereby the "optimal" value of N is estimated in advance, to the solution of our problem. In order to simplify our notation we divide each equation of (4.12b) through by y and then define a sequence $v_n^{[N]}$ by the relation

$$v_n^{[N]} = \frac{y^n}{y}, \qquad n = 1, 2, \ldots, N-1.$$

System (4.12b) may now be written in the form

$$\mathbf{A}\mathbf{v}^{[N]} = \mathbf{b},$$

where $\mathbf{A}$ is the coefficient matrix of (4.12b), $\mathbf{v}^{[N]} = (v_1^{[N]}, v_2^{[N]}, \ldots, v_{N-1}^{[N]})^T$ and $\mathbf{b} = (-c_{1,2}(x), -c_{2,3}(x), \ldots, -c_{N-1,N}(x))^T$. We may now use simple forward elimination to annihilate all of the coefficients appearing below the main diagonal of the matrix $\mathbf{A}$ to produce a sequence of relations of the form

$$p_{r+1}v_r^{[N]} - v_{r+1}^{[N]} = e_r, \qquad r = 1, 2, \ldots, N-1.$$

Replacing N by $N+1$ we have

$$p_{r+1}v_r^{[N+1]} - v_{r+1}^{[N+1]} = e_r$$

and subtracting these two relations we obtain

$$\begin{aligned} v_r^{[N+1]} - v_r^{[N]} &= \frac{1}{p_{r+1}}[v_{r+1}^{[N+1]} - v_{r+1}^{[N]}] \\ &= \frac{1}{p_{r+1}p_{r+2}}[v_{r+2}^{[N+1]} - v_{r+2}^{[N]}] \\ &\cdots\cdots\cdots\cdots\cdots\cdots \\ &\cdots\cdots\cdots\cdots\cdots\cdots \\ &= \frac{1}{p_{r+1}p_{r+2}\cdots p_N}[v_N^{[N+1]} - v_N^{[N]}] \\ &= \frac{e_N}{\prod_{i=r+1}^{N+1} p_i}. \end{aligned}$$

This relation is the analogue of relation (4.10) appearing in Olver's paper (Olver. 1967). Our procedure from now on is generally simpler than that which would be used for recurrence relations since we are only interested in convergence to $v_1^{[N]}$ and we are not concerned with any other members of the vector $\mathbf{v}^{[N]}$. Thus if we require convergence to D decimal places we continue our elimination process for increasing values of r until

$$\left| e_r \Big/ \prod_{i=2}^{r+1} p_i \right| < \tfrac{1}{2} \times 10^{-D},$$

and having found the first value of r for which this inequality holds we set $N = r$. It can be seen that this whole procedure follows very closely the algorithm derived by Olver for the solution of linear second order recurrence relations. Having obtained convergence to a limiting value $v_1^{[r]}$ which is such that $|v_1^{[r+1]} - v_1^{[r]}| < \frac{1}{2} \times 10^{-D}$ it finally remains to integrate the first order differential equation

$$y' = v_1^{[r+1]}y$$

and this can be done using any standard discretisation algorithm given one initial condition of the form $y(x_0) = y_0$. It must be emphasised at this stage that any accuracy requirements will usually be specified on the solution $y(x)$ rather than on the ratio y'/y. In view of this it will normally be necessary to integrate both of the equations

$$y' = v_1^{[r]}y \qquad \text{and} \qquad y' = v_1^{[r+1]}y$$

to make sure that the two solutions $y^{[r]}$ and $y^{[r+1]}$ of these equations do differ by less than D decimal places at each steppoint. If this is not the case it will be necessary to compute $v_1^{[r+2]}$ and to integrate the system

$$y' = v_1^{[r+2]}y$$

for the solution $y^{[r+2]}$. This procedure will be continued until two solution sequences $y^{[r+j]}$ and $y^{[r+j+1]}$ differ by less than the prescribed amount. To guard against the likelihood of having to compute $v_1^{[r+2]}$ it is advisable, of the solution $y(x)$ is required correct to D decimal places, to carry out the forward elimination process until a value of r is found for which

$$\left| e_r \Big/ \prod_{i=2}^{r+1} p_i \right| < \tfrac{1}{2} \times 10^{-(D+1)},$$

i.e. to carry one extra guarding figure.

Thus by identifying a variable $v_i^{[N]}$ with the derivative quotient y^i/y we may extend Olver's algorithm in a natural way to the numerical solution of linear second order ordinary differential equations. By carrying out an analysis of the truncation error of our scheme similar to that described by Olver (1967), section 5, we can show that the intermediate solution $v_1^{[r]}$ obtained using this process is indeed the required solution to this part of our problem. We shall return to this technique at a later stage but for the time being we consider a numerical example which serves to illustrate some of the algorithms which have been derived in this chapter so far.

Example 4.1

As a rather artificial but nevertheless illuminating example we consider the integration of the equation

$$y'' - 100y' + 99y = 0, \tag{4.13a}$$

where the prescribed boundary conditions take the form

$$y(0) = 1,$$

$$|y''/y| \quad \text{is relatively "small" for all } x > 0.$$

The general solution of (4.13a) is given by

$$y(x) = c_1 e^x + c_2 e^{99x}$$

and it is clear that the boundary conditions are such that $c_1 = 1$, $c_2 = 0$. Applying iteration scheme (4.6) to the solution of (4.13a) we obtain

$$(y''/y)^{(t-1)} - 100(y'/y)^{(t)} + 99 = 0. \tag{4.13b}$$

Putting $(y''/y)^{(0)} \equiv 0$ initially we obtain

$$(y'/y)^{(1)} = \tfrac{99}{100} = A_0. \tag{1}$$

Differentiating this relation formally with respect to x we obtain $(y''/y)^{(1)} = (\tfrac{99}{100})^2$ and substituting this expression into (4.13b) for $t = 2$ we obtain

$$\begin{aligned}(y'/y)^{(2)} &= \frac{99}{100} + \frac{1}{100}\left(\frac{99}{100}\right)^2 \\ &= \frac{999\,801}{10^6}.\end{aligned} \tag{2}$$

If, however, we adopt the second approach defined by (4.11) we have

$$\begin{aligned}\hat{A}_2 &= \frac{999801}{10^6} - \frac{2 \times \dfrac{99}{100}\left[\dfrac{999801}{10^6} - \dfrac{99}{100}\right]}{-100 + 2 \times \dfrac{99}{100}}, \\ &= \frac{999801}{10^6} + \frac{198}{10^6} \times \frac{9801}{9802}, \\ &\approx \frac{999999}{10^6}.\end{aligned}$$

This illustrates the rapid convergence of our modified algorithm compared with that of our original algorithm for this particular problem. It is clear that both of these iteration schemes are converging rapidly to the limit $y'/y = 1$ which is the required solution. We now consider a second method of approach based on Miller's backward recurrence algorithm. Relation (4.13a) and its first and second derivatives with respect to x are given by

$$y'' - 100y' + 99y = 0,$$

$$y''' - 100y'' + 99y' = 0,$$

$$y'''' - 100y''' + 99y'' = 0.$$

It can be seen that these relations take a particularly simple form due to the fact that (4.13a) has constant coefficients. Setting $N = 2, 3, 4$ in turn we obtain

$$y' = \frac{99}{100}y, \quad y' = \frac{9900}{9901}y \quad \text{and} \quad y' = \frac{980199}{980200}y$$

respectively. It can be seen from these expressions that the algorithm is converging rapidly to the required solution. The need to choose the initial sequence $(y''/y)^{(0)}$ to be everywhere "small" initially when using iteration scheme (4.6) is highlighted by the fact that if we choose $(y''/y)^{(0)} = 99^2$ we obtain convergence to the unwanted solution e^{99x} and if we choose $(y''/y)^{(0)} > 99^2$ then the algorithm does not converge at all. In general if any of our algorithms do converge then they do so to a relation of the form

$$(y'/y)^{(t)} = A_t(x).$$

Providing we now use a stable integration procedure with a sufficiently small value of the steplength of integration to integrate our final first order ordinary differential equations then convergence to the required solution is assured.

4.3 Computational aspects

We now consider some important computational aspects concerning the practical implementation of the algorithms which have been discussed in this chapter. We consider first the iterative algorithm defined by relations (4.8). Firstly we note that at no stage of our algorithm do we need to know explicit expressions for the intermediate values $y^{(t)}$ and so only one first order ordinary differential equation needs to be integrated, namely the equation obtained on the convergence of the sequence A_i. Secondly we note that our algorithms are based on successive differentiation which is known to be a rather unreliable scheme numerically. This problem may, however, be completely avoided if an algebra system is used to perform all differentiations analytically (or of course if differentiations are performed by hand). The algorithms described earlier in this section have been implemented using the algebra system CAMAL (Barton *et al.*, 1970) and they have proved to be generally more reliable than algorithms incorporating numerical differentiation. A further point to note is that so far we have placed no restriction on the solution $y_1(x)$ except to demand that condition (4.5) should hold. This condition may be satisfied even when $y_1(x)$ is a rapidly increasing exponential function. In cases such as this the original initial value problem (4.2), as it is usually posed with values of y and y' given at the initial point, is ill-conditioned. An examination of Example 4.1 reveals that even though this particular problem would be very badly conditioned if posed in the usual way, and so cannot be integrated effectively using a standard finite difference method, our iterative procedures perform quite satisfactorily and in fact the more badly posed our problem is, the better our iterative algorithms generally perform. Thus by adopting this iterative approach we are effectively transforming a badly posed problem into a well posed one essentially by changing the boundary conditions which we use to define the required solution. Exactly why this phenomenon occurs can be most clearly illustrated by considering some particular boundary conditions. If we consider the integration of (4.13a) with initial conditions $y(0) = y'(0) = 1$, it is clear that if we change the initial conditions very slightly the solution is changed by a great deal causing the problem to be ill-conditioned. If, however, we use the conditions $y(0) = 1$, $|y''/y|$ is relatively small everywhere, we see that the second condition ensures that the solution which we obtain is directly proportional to e^x. It is obvious, but perhaps worth stating explicitly, that this second condition cannot be perturbed by a small amount (!) and if the first condition is changed to $\tilde{y}(0) = 1 + \varepsilon$ the corresponding solution is $\tilde{y}(x) = (1 + \varepsilon)e^x$; i.e. there is a small relative change in the solution obtained. Finally we note that there is a set of miscellaneous computational procedures which we still have to consider, such as exactly how we test for the convergence of our algorithms, how we actually integrate our final system of ordinary differential equations, etc. Before considering these aspects of our algorithm, however, we first derive an algorithm for the solution of a special class of second order ordinary differential equations.

4.4 A special class of second order O.D.E.s

The problem which we now wish to consider is the numerical integration of the homogeneous second order ordinary differential equation

$$y'' - f(x)y = 0, \tag{4.14}$$

where we assume that the function $f(x)$ is large and positive throughout the region of integration. We further assume that it is possible to choose two complementary functions $y_1(x)$ and $y_2(x)$ of (4.14) which are linearly independent and which are such that

$$\lim_{x\to\infty} \frac{y_2(x)}{y_1(x)} = 0.$$

It is clear that the iterative algorithms which we have so far considered in this chapter are not applicable to the solution of (4.14) as it stands due to the absence of a y' term. For the time being we shall not be concerned with which solution of (4.14) we require but we will instead attempt to reduce equation (4.14) to one of the form (4.2) by making a transformation of the form

$$y(x) = z(x)\mathrm{e}^{p(x)}$$

(see also Cash, 1975). Substituting this expression into (4.14) we obtain

$$\{z''(x) + 2p'(x)z'(x) + [p'(x)]^2 z(x) + p''(x)z(x)\}\mathrm{e}^{p(x)}$$
$$- f(x)z(x)\mathrm{e}^{p(x)} = 0. \qquad (4.15)$$

What we shall be attempting to do is to generate the required solution of this equation using one of the iterative schemes described earlier in this chapter. As a result we require the coefficient of $z'(x)$ to be large throughout the range of integration. Suppose now, with this in mind, we choose,

$$[p'(x)]^2 = f(x),$$

since $f(x)$ is large by assumption; then equation (4.15) reduces to

$$z''(x) + 2p'(x)z'(x) + p''(x)z(x) = 0 \qquad (4.16)$$

with $p(x) = \pm\int_0^x [f(\bar{x})]^{\frac{1}{2}}\,\mathrm{d}\bar{x}$. One possible way of generating the required solution $z(x)$ of equation (4.16) is to use the iteration scheme

$$\left(\frac{z''}{z}\right)^{(t-1)} + 2p'(x)\left(\frac{z'}{z}\right)^{(t)} + p''(x) = 0,$$
$$z''^{(0)} \equiv 0, \qquad (4.17)$$

where the particular choice for the function $p(x)$ depends on the particular solution of (4.14) which we require. If we require the dominant solution we take $p(x)$ to be positive whereas if we require the other solution we choose $p(x)$ to be negative. Another possible way of generating $z(x)$ is to use the extension of Olver's algorithm described earlier in this chapter with the same rules for choosing $p(x)$ again applying. One particular application of this approach, which we use to describe the computational aspects of our algorithms, is in the tabulation of *Airy functions* which are solutions of the equation

$$y'' - xy = 0.$$

It turns out that, as the value of x increases, this equation becomes increasingly difficult to integrate using standard methods due to the large local truncation error associated with the unwanted solution. This generally forces the steplength of integration to become excessively small. If, however, we use the algorithms described earlier in this chapter we find that, the larger the value of x is, the more quickly our algorithms converge. In view of this a sensible procedure to adopt in practice when integrating equations of this type (i.e. when $f(x)$ is monotonically increasing and becomes large for large x) would be to use a standard integration procedure for small values of x and to switch to one of our iterative algorithms when x becomes "sufficiently large". A similar approach may be developed in the case where $f(x)$ is large and negative. In this case we make the substitution

$$y(x) = z(x)\,[\cos p(x) + \sin q(x)], \tag{4.18}$$

where p and q are at this stage arbitrary functions of x. Formally differentiating relation (4.18) with respect to x we obtain

$$\begin{aligned}
y'(x) = {} & z'(x)\,[\cos p(x) + \sin q(x)] \\
& + z(x)[-p'(x)\sin p(x) + q'(x)\cos q(x)], \\
y''(x) = {} & z''(x)[\cos p(x) + \sin q(x)] \\
& + 2z'(x)[-p'(x)\sin p(x) + q'(x)\cos q(x)] \\
& + z(x)[-p''(x)\sin p(x) + q''(x)\cos q(x)] \\
& - z(x)[[p'(x)]^2\cos p(x) + [q'(x)]^2\sin q(x)].
\end{aligned}$$

Substituting this latter relation into (4.14) we obtain a linear second order ordinary differential equation for the unknown solution $z(x)$. If we now choose

$$[p'(x)]^2 = [q'(x)]^2 = -f(x)$$

this equation simplifies to

$$\frac{z''(x)}{z(x)}\{\cos p(x) + \sin p(x)\}$$

$$+ 2\frac{z'(x)}{z(x)}\{-p'(x)\sin p(x) + p'(x)\cos p(x)\}$$

$$+ \{-p''(x)\sin p(x) + p''(x)\cos p(x)\} = 0 \tag{4.19}$$

and we may again seek to generate the required solution of this equation using one of the techniques described earlier. Special care has to be taken when either the coefficient of $z'(x)$ is zero (i.e. when $\sin p(x) = \cos p(x)$) or when $z(x)$ is zero. The case where $z(x)$ becomes zero at a finite set of points $\{x_i\}$ has already been considered in discussing the solution of (4.3) using (4.6). Basically what we do is to use a standard integration procedure to carry out the integration in a small neighbourhood of each x_i and then we use our iterative scheme outside these small neighbourhoods. Exactly the same procedure may also be used in the neighbourhood of points for which $\sin p(x) = \cos p(x)$ and so in general this case causes no real computational problems.

4.5 A particular example

In order to explain in more detail the computational aspects of our algorithms we consider the integration of *Airy's equation*

$$y'' - xy = 0 \tag{4.20}$$

in a region where x is large and positive. This equation has two linearly independent solutions, usually denoted by $\mathbf{Ai(x)}$ and $\mathbf{Bi(x)}$, which have the integral representations

$$\mathbf{Ai(x)} = \frac{1}{\pi}\int_0^\infty \cos(\tfrac{1}{3}t^3 + xt)\mathrm{d}t$$

and

$$\mathbf{Bi(x)} = \frac{1}{\pi}\int_0^\infty \{\exp(-\tfrac{1}{3}t^3 + xt) + \sin(\tfrac{1}{3}t^3 + xt)\}\,\mathrm{d}t.$$

It may be shown that for large x

$$\mathbf{Bi(x)} \to \infty \qquad \text{and} \qquad \mathbf{Ai(x)} \to 0.$$

As the value of x increases, the integration of equation (4.20) for the recessive solution $\mathbf{Ai(x)}$ using standard finite difference integration procedures becomes increasingly inefficient. This is due to the excessively small steplength of integration which needs to be used in order to ensure that the dominant solution **Bi(x)** is adequately represented by the discretisation scheme. Although our problem may in no way be regarded as being stiff there are some similarities between the problem of computing **Ai(x)** and the problem encountered when integrating certain stiff systems in that it is an unwanted solution which determines the steplength of integration that is to be used. If we now make the transformation $y(x) = z(x)e^{p(x)}$ and substitute for $y(x)$ in (4.20) we obtain the relation (4.16) where $p(x) = -\frac{2}{3}x^{3/2}$. Iteration scheme (4.17) now becomes

$$\left(\frac{z''}{z}\right)^{(t-1)} - 2x^{\frac{1}{2}}\left(\frac{z'}{z}\right)^{(t)} - \tfrac{1}{2}x^{-\frac{1}{2}} = 0. \tag{4.21a}$$

For the purposes of our example we assume that we have integrated forward for **Ai (x)** up to the steppoint $x = 10$ using a standard discretisation scheme and that we have found that as x increases our scheme has become increasingly inefficient due to the small step that is required to keep the local truncation error sufficiently small. We further assume that we have decided to use instead the iteration scheme (4.21a) from the point $x = 10$ onwards. In order to use scheme (4.21a) we make the initial approximation $(z''/z)^{(0)} \equiv 0$. The sequence of iterates obtained from (4.21a) are

$$(z'/z)^{(1)} = -\frac{1}{4x}$$

$$(z'/z)^{(2)} = -\frac{1}{4x} + \frac{5}{32x^{5/2}}$$

$$(z'/z)^{(3)} = -\frac{1}{4x} + \frac{5}{32x^{5/2}} - \frac{30}{128x^4} + \frac{25}{2048x^{11/2}}$$

. . . .

. . . .

In our numerical example we shall assume that we wish to compute the solution $z(x)$ correct to four decimal places at steppoints a distance 0·5 apart. We note that each iteration takes the form $(z'/z)^{(t)} = A_t(x)$ where the sequence $A_t(x)$ is defined recursively by (4.8). Iterations are continued until

$$|A_t(10) - A_{t-1}(10)| < \tfrac{1}{2} \times 10^{-4} \tag{4.21b}$$

for some value of t, i.e. until successive values of z'/z differ by less than $\frac{1}{2} \times 10^{-4}$. When this is the case the relation $z'(10) = A_t(10)z(10)$ is taken as the finally accepted "reduced" differential equation at the point $x = 10$. In our particular example it was found that the inequality (4.21b) was satisfied when $t = 3$. Our next problem is to test whether

$$|A_3(10\cdot5) - A_2(10\cdot5)| < \tfrac{1}{2} \times 10^{-4}$$

and if it is we take $z'(10\cdot5) = A_3(10\cdot5)z(10\cdot5)$ as our finally accepted reduced differential equation at the point $x = 10\cdot5$. If this inequality is not satisfied we need to perform further iterations at this point until a value of t is found for which

$$|A_t(10\cdot5) - A_{t-1}(10\cdot5)| < \tfrac{1}{2} \times 10^{-4}.$$

A similar procedure may be used at all steppoints in the range of integration. It was in fact found that $|A_3(x) - A_2(x)| < \frac{1}{2} \times 10^{-4}$ for all $x \geq 10$ and so it was not necessary to compute any additional iterates past $A_3(x)$. Our main outstanding problem now is to integrate the final sequence of reduced first order ordinary differential equations for $z(x)$. Since we require the solution at intervals of 0·5 in the argument x and since we are not expecting our final system of O.D.E.s to be stiff we shall perform our integration with a step $h = 0\cdot5$ using an Adams predictor–corrector scheme used in PECE mode as suggested by Shampine and Gordon (1975). For our purposes we use a first order predictor–corrector pair consisting of the explicit Euler rule as predictor and the trapezoidal rule as corrector but a higher order pair may be used if this does not give sufficient precision. The actual way in which this predictor–corrector pair may be implemented has been fully described by Shampine and Gordon (1975) and we follow their approach very closely. In particular the difference between the predicted and the corrected solutions will serve as an estimate of the local truncation error in the predicted solution. If this estimate is less than a prescribed tolerance, local extrapolation will be performed and the corrected solution will be the one which is actually carried forward. If this first order predictor–corrector pair is used to integrate our final sequence of first order ordinary differential equations our overall procedure will be self-starting. If, however, we wish to use a higher order Adams predictor–corrector pair it will be necessary for starting purposes to first of all compute some "back values" and this may be done by applying the algorithm at some previous steppoints where the solution has already been computed (i.e. at $x = 9\cdot5, 9\cdot0, \ldots$).

The results obtained for the integration of our problem are given in Table 4.1 and as can be seen the agreement with the true solution $z(x)$ is satisfactory. In order to find the final solution $y(x)$ we merely need to use the relation

Table 4.1

x	Predicted value of z	Corrected Value of z	True value of z
10·5	0·15616	0·15622	0·15622
11·0	0·15439	0·15444	0·15444
11·5	0·15271	0·15276	0·15276
12·0	0·15112	0·15117	0·15117
13·0	0·14818	0·14822	0·14822
14·0	0·14549	0·14552	0·14553
15·0	0·14304	0·14306	0·14307
16·0	0·14077	0·14079	0·14080
17·0	0·13868	0·13869	0·13870

$y(x) = z(x)e^{-(2x^{3/2}/3)}$. We may also use the problem of integrating equation (4.20) for the recessive solution **Ai(x)** to illustrate the extension of Olver's algorithm to the solution of second order ordinary differential equations. In this case relation (4.16) becomes

$$z'' - 2x^{\frac{1}{2}}z' - \tfrac{1}{2}x^{-\frac{1}{2}}z = 0.$$

Formally differentiating this relation with respect to x we obtain

$$z''' - 2x^{\frac{1}{2}}z'' - \tfrac{3}{2}x^{-\frac{1}{2}}z' + \tfrac{1}{4}x^{-\frac{3}{2}}z = 0$$

$$z'''' - 2x^{\frac{1}{2}}z''' - \tfrac{5}{2}x^{-\frac{1}{2}}z'' + x^{-\frac{3}{2}}z' - \tfrac{3}{8}x^{-\frac{5}{2}}z = 0$$

$$z''''' - 2x^{\frac{1}{2}}z'''' - \tfrac{7}{2}x^{-\frac{1}{2}}z''' + \tfrac{9}{4}x^{-\frac{3}{2}}z'' - \tfrac{15}{8}x^{-\frac{5}{2}}z' + \tfrac{15}{16}x^{-\frac{7}{2}}z = 0$$

$$\cdots\cdots\cdots\cdots\cdots$$

By carrying out this successive differentiation we may build up the system of differential equations defined by (4.12). We now carry out simple forward elimination to obtain the two sequences $\{p_r\}$ and $\{e_r\}$ defined earlier. This procedure is continued until

$$\left|\frac{e_r(10)}{p_2(10)p_3(10)\ldots p_{r+1}(10)}\right| < \frac{1}{2} \times 10^{-4}$$

and when a value of r is found for which this inequality holds we set $N = r$. Having done this we apply exactly the same procedure at $x = 10{\cdot}5$ and if

$$\left|\frac{e_N(10{\cdot}5)}{p_2(10{\cdot}5)p_3(10{\cdot}5)\ldots p_{N+1}(10{\cdot}5)}\right| < \frac{1}{2} \times 10^{-4} \tag{4.21c}$$

we keep $N = r$ and carry out back substitution to find the ratio z'/z. If inequality (4.21c) does not hold it will be necessary to test the validity of this inequality for $r = N + 1, N + 2, \ldots$ until (4.21c) is again satisfied. For our problem it is found that (4.21c) is satisfied with $N = 4$ at all points in the range of integration and the results obtained are given in Table 4.2. As can be seen satisfactory agreement with the true solution is again obtained.

TABLE 4.2
Value of $N = 4$

x	Value of Z
10·5	0·15622
11·0	0·15444
11·5	0·15276
12·0	0·15117
13·0	0·14821

As a final comment on our algorithms for the solution of linear second order ordinary differential equations we note that there is a relationship between the iterative algorithm defined by relation (4.6) and the classical *Riccati transformation* method. The Riccati scheme involves making the transformation $y' = A(x)y$ initially thus reducing the problem to one of finding the function $A(x)$. Substituting this transformation into (4.2) we obtain the non-linear first order ordinary differential equation

$$A'(x) + A^2(x) + A(x)p(x) + q(x) = 0.$$

It may now well be that this equation is non-stiff (as it is in the case of Example 4.1) and so may be integrated in a satisfactory manner using a standard explicit method. A major difference between the two approaches, however, is that the Riccati transformation method, as it is usually applied, is limited to linear differential equations whereas, as we shall see later in this chapter, our iterative algorithms are specially designed so that they may be applied to a large class of non-linear equations as well.

4.6 Ordinary differential equations of order greater than 2

We now consider fairly briefly the extension of our techniques to the solution of higher order linear ordinary differential equations. For the purposes of our discussion we shall confine our attention to the third order equation

$$y''' + p(x)y'' + q(x)y' + r(x)y = 0 \tag{4.22}$$

since the extension to even higher order linear equations is relatively straightforward once this case is well understood. The general solution, $y(x)$, of equation (4.22) may be written in the form

$$y(x) = c_1y_1(x) + c_2y_2(x) + c_3y_3(x), \tag{4.23}$$

where $y_1(x)$, $y_2(x)$ and $y_3(x)$ are any three solutions of (4.22) which are such that at some point x_0 in the range of integration the vectors $[y_1(x_0), y_1'(x_0), y_1''(x_0)]^T$, $[y_2(x_0), y_2'(x_0), y_2''(x_0)]^T$ and $[y_3(x_0), y_3'(x_0), y_3''(x_0)]^T$ are linearly independent, and the c_i are arbitrary constants. We assume first of all that it is possible to choose the basis solutions $y_i(x)$ appearing in (4.23) such that

$$\left|\frac{y_i'''(x)}{y_i(x)}\right| \ll \left|\frac{y_1'''(x)}{y_1(x)}\right|, \qquad i = 2,3,$$

and that we wish to compute a solution of (4.22) which does not contain a component of $y_1(x)$. Following the procedure adopted in the case of linear second order equations we attempt to replace (4.22) by an "equivalent" well conditioned boundary value problem by reducing it to a linear second order equation which does not contain a component of $y_1(x)$ but which has $y_2(x)$ and $y_3(x)$ as a basis for its general solution. We assume that associated with (4.22) we are given two initial conditions of the form $y(x_0) = y_0$ and $y'(x_0) = y_0'$ together with the additional information that the required solution is of the form (4.23) with $c_1 = 0$; i.e. we want a solution with small relative variation. We shall attempt to generate the required solution of (4.22) by first of all extracting from it a "quadratic factor" using an approach similar to that described in Chapter 2. Suppose now that the quadratic factor which we require has the form

$$y'' = A(x)y' + B(x)y.$$

Substituting this into (4.22) we obtain the relation

$$[A'(x) + \{A(x)\}^2 + B(x) + p(x)A(x) + q(x)]y'(x)$$
$$+ [A(x)B(x) + B'(x) + p(x)B(x) + r(x)]y(x) = 0.$$

Since this relation is valid for all solutions of our quadratic factor it follows that

$$\left.\begin{aligned} A'(x) + \{A(x)\}^2 + B(x) + p(x)A(x) + q(x) = 0 \\ A(x)B(x) + B'(x) + p(x)B(x) + r(x) = 0 \end{aligned}\right\} \tag{4.24}$$

and we may generate the required solution of this system using one of the iteration schemes derived in Chapter 2. The particular iteration scheme which we consider for the solution of our problem is

$$A_t(x) = -\left\{\frac{q(x)}{p(x)} + \frac{1}{p(x)}[A'_{t-1}(x) + \{A_{t-1}(x)\}^2 + B_{t-1}(x)]\right\}, \quad t = 2, 3, \ldots$$

$$B_t(x) = -\left\{\frac{r(x)}{p(x)} + \frac{1}{p(x)}[A_{t-1}(x)B_{t-1}(x) + B'_{t-1}(x)]\right\}$$

with $A_1(x) = -\dfrac{q(x)}{p(x)}, B_1(x) = -\dfrac{r(x)}{p(x)}$. We again see that this algorithm involves extensive differentiation and so may be used to advantage with an algebra system. This iteration scheme may be applied in the usual way until successive values of A_t and successive values of B_t both differ by less than a prescribed amount (assuming of course that convergence does occur). When this is the case the final second order ordinary differential equation may be integrated using a standard method using the two given initial conditions $y(x_0) = y_0, y'(x_0) = y'_0$. Useful sufficient conditions for the convergence of this scheme are again difficult to derive but a partial analysis may be found in Cash (1972). This iteration scheme which we have just described is one of several possible schemes for the solution of (4.24). Another scheme may be derived by setting $A'_0(x) = B'_0(x) = 0$ initially. Equations (4.24) now define two simultaneous non-linear algebraic equations for $A_1(x)$ and $B_1(x)$ and a solution of these may be generated using Newton's method. This approach may be continued iteratively for increasing integer values of t. In general this scheme will require more computational effort than the one which we have just described but it is found to converge rather faster.

We now mention an extension of the Miller/Olver type of approach to deal with the solution of this problem. Equation (4.22) and its successive derivatives are given by

$$y''' + p(x)y'' + q(x)y' + r(x)y = 0$$

$$y'''' + p(x)y''' + [p'(x) + q(x)]y'' + [q'(x) + r(x)]y' + r'(x)y = 0$$

$$y''''' + p(x)y'''' + [2p'(x) + q(x)]y''' + [p''(x) + 2q'(x) + r(x)]y'' + [q''(x) + 2r'(x)]y' + r''(x)y = 0$$

. . . .

. . . .

The mth equation of this system may be written in the form

$$y^{m+2} + \sum_{i=0}^{m+1} c_{m,i}(x) y^i = 0.$$

Writing this linear system in matrix form we have

$$\begin{bmatrix} c_{1,2}(x) & 1 & & & \\ c_{2,2}(x) & c_{2,3}(x) & 1 & & \\ c_{3,2}(x) & c_{3,3}(x) & c_{3,4}(x) & 1 & \\ \cdot & & & & \\ \cdot & & & & \\ \cdot & & & & \\ c_{N-2,2}(x) & c_{N-2,3}(x) & c_{N-2,4}(x) & \cdots\cdots & c_{N-2,N-1}(x) \end{bmatrix} \begin{bmatrix} y'' \\ y'''' \\ y'''' \\ \cdot \\ \cdot \\ \cdot \\ y^{N-1} \end{bmatrix}$$

$$= \begin{bmatrix} -c_{1,1}(x) \\ -c_{2,1}(x) \\ -c_{3,1}(x) \\ \cdot \\ \cdot \\ \cdot \\ -c_{N-2,1}(x) \end{bmatrix} y' + \begin{bmatrix} -c_{1,0}(x) \\ -c_{2,0}(x) \\ -c_{3,0}(x) \\ \cdot \\ \cdot \\ \cdot \\ -c_{N-2,0}(x) \end{bmatrix} y. \tag{4.25}$$

Following the approach adopted by Miller with his backward recurrence algorithm we now set $y^{m+1} = y^m = 0$ for some large value of $m = N$. This produces $N - 2$ equations in the $N - 2$ unknowns $y'', y''', \ldots, y^{N-1}$. By eliminating these variables in the direction of decreasing m we obtain a linear expression for y'' in terms of y' and y. We then integrate this second order equation using a standard technique. As an alternative we may implement Olver's approach which attempts to estimate an "optimal" value of N before any solution values have been computed. By eliminating all the elements below the main diagonal of the coefficient matrix of (4.25) we obtain a sequence of relations of the form

$$p_{r+1} y_N^r - y_N^{r+1} = e_r y' + g_r y,$$

$$r = 2, 3, \ldots, N-2, \quad \text{where} \quad y_N^r = \frac{d^r y_N}{dx^r}.$$

Replacing N by $N + 1$ this relation becomes

$$p_{r+1}y_{N+1}^{r} - y_{N+1}^{r+1} = e_r y' + g_r y.$$

Subtracting these two relations we obtain

$$y_{N+1}^{r} - y_N^{r} = \frac{1}{p_{r+1}}[y_{N+1}^{r+1} - y_N^{r+1}]$$

$$= \frac{1}{p_{r+1}p_{r+2}}[y_{N+1}^{r+2} - y_N^{r+2}]$$

$$\cdots\cdots\cdots\cdots$$

$$= \frac{y_{N+1}^{N}}{\prod_{i=r+1}^{N} p_i}$$

$$= \frac{e_N y' + g_N y}{\prod_{i=r+1}^{N+1} p_i}. \tag{4.26}$$

Exactly the same procedure as was used earlier with second order equations may now be applied. In particular we are only interested in obtaining convergence to y'' and are not concerned with any y_N^r for $r > 2$. If we require convergence to D decimal places we continue our elimination process for increasing values of r until the two inequalities

$$\left| e_r \Big/ \prod_{i=3}^{r+1} p_i \right| < \tfrac{1}{2} \times 10^{-D} \text{ and } \left| g_r \Big/ \prod_{i=3}^{r+1} p_i \right| < \tfrac{1}{2} \times 10^{-D}$$

both hold and having found the first value of r for which these inequalities are valid we set $N = r$. Having obtained the final converged value for y'' it remains to integrate a linear second order O.D.E. of the form

$$y'' = \bar{e}_N y' + \bar{g}_N y$$

and this may be done using a standard method. Since the accuracy requirement initially specified will usually be imposed on the required solution $y(x)$ rather than on the coefficients of our reduced equation it will be necessary to integrate a linear second order O.D.E. of the form

$$y'' = \bar{e}_N y' + \bar{g}_N y$$

and

$$y'' = \bar{e}_{N-1}y' + \bar{g}_{N-1}y.$$

This is to check that the final solution sequences do indeed differ by less than the prescribed amount in some appropriate sense. If this is not the case it will be necessary to continue the process by computing

$$y'' = \bar{e}_i y' + \bar{g}_i y, \qquad i = N+1, N+2, \ldots,$$

until a value of i is found for which the final solution values are sufficiently close.

Finally in this section we consider the case where

$$\left|\frac{y_1'''(x)}{y_1(x)}\right| \quad \text{and} \quad \left|\frac{y_2'''(x)}{y_2(x)}\right| \text{ are both} \gg \left|\frac{y_3'''(x)}{y_3(x)}\right|$$

and we wish to compute a solution of (4.22) which is directly proportional to $y_3(x)$. The boundary conditions which we use for the solution of this problem are $y(x_0) = y_0$ and the knowledge that the required solution is of the form (4.23) with $c_1 = c_2 = 0$. Following the procedure adopted in the analogous case for linear recurrence relations we could seek to generate a solution of our problem using the iteration scheme

$$\left(\frac{y'}{y}\right)^{(t)} = -\frac{r(x)}{q(x)} - \frac{p(x)}{q(x)}\left(\frac{y''}{y}\right)^{(t-1)} - \frac{1}{q(x)}\left(\frac{y'''}{y}\right)^{(t-1)}. \tag{4.27}$$

By so doing we reduce our original third order differential equation to one of the first order in the quantity $y^{(t)}$. In order to use scheme (4.27) we first set $(y''/y)^{(0)} = (y'''/y)^{(0)} = 0$. We note that any sequence of initial approximations could be used to get iteration scheme (4.27) started but as usual we choose all higher derivatives of y to be zero initially to reflect the fact that we are seeking a solution with small relative variation. Successively applying relation (4.27) for increasing values of the integer t we obtain a sequence of relations of the form

$$(y'/y)^{(t)} = A_t(x),$$

where

$$A_1(x) = -\frac{r(x)}{q(x)}$$

and

$$A_t(x) = A_1(x) - \frac{1}{q(x)}\{p(x)[\{A_{t-1}(x)\}^2 + A'_{t-1}(x)] + \{A_{t-1}(x)\}^3 + 3A_{t-1}(x)A'_{t-1}(x) + A''_{t-1}(x)\}, t = 2, 3, \ldots$$

Assuming that this scheme does converge for increasing t, in the sense that there exists an N_ε such that

$$|A_{N_\varepsilon}(x) - A_{N_\varepsilon - 1}(x)| < \varepsilon$$

for any given ε, we may integrate the final first order ordinary differential equation $y' = A_{N_\varepsilon}(x)y$ using a standard integration procedure together with the initial condition $y(x_0) = y_0$. We shall not consider this scheme any further since it is a straightforward extension of those considered earlier.

We now consider the extension of some of the techniques derived earlier in this chapter to produce algorithms suitable for the approximate numerical integration of the first order system of ordinary differential equations

$$\mathbf{y}' = \mathbf{f}(x, \mathbf{y}), \quad \mathbf{y}(x_0) = \mathbf{y}_0. \tag{4.28}$$

As we shall see, the basic algorithm which we propose for the integration of (4.28) is a relatively straightforward extension of algorithms considered earlier in this chapter and it is the implementation and computational aspects of our extended algorithm which require careful consideration. In view of this the next two sections will be largely descriptive. For our purposes we shall find it convenient to consider the linear and non-linear cases separately and correspondingly we devote the next section to an examination of the linear case.

4.7 Linear first order systems of ordinary differential equations

In this section we shall be concerned with the approximate numerical integration of the first order system

$$\frac{d\mathbf{y}}{dx} = \mathbf{A}(x)\mathbf{y}, \qquad \mathbf{y}(x_0) = \tilde{\mathbf{y}}_0 \tag{4.29}$$

in the finite range $x_0 \leq x \leq x_N$, where $\mathbf{y}$ and $\tilde{\mathbf{y}}_0$ are s-dimensional vectors and $\mathbf{A}$ is an $s \times s$ matrix. The algorithm which we shall describe has been developed for use with a particular explicit integration procedure (namely Taylor series), and the algorithm will be described with this particular scheme in mind. However, it will soon become clear that most other automatic integration procedures may be used instead and that Taylor series is, in fact, only used for demonstration purposes. Up to now the method of Taylor series has not been used very widely in practice for the numerical integration of either stiff or non-stiff systems of equations due mainly to the fact that it is often extremely difficult to compute the higher derivatives of $\mathbf{y}$ by hand. This is true even for moderately sized systems. Due to the advent of manipulative programs (Barton *et al.*, 1971a, b), which allow the relevant Taylor series to be generated automatically, it is now possible to integrate quite large systems of equations using the Taylor series method with the programmer having to do little more than merely "present" his system of equations to the automatic package. Our first major difficulty, when faced with the problem of integrating (4.29) numerically, is to decide when we shall regard this system as being stiff so that the straightforward integration of (4.29) using an explicit method becomes inefficient. As a result of experiments performed with the *Cambridge Taylor Series System* (Barton *et al.*, 1971b), which is the particular version of Taylor series on which our algorithm is based and which we shall refer to as CTSS from now on, it seems convenient to regard a system as being stiff if it has a stiffness ratio greater than 10. This conclusion is based on experiments performed on an uncoupled 2×2 system of the form (4.29) with solutions e^{-x} and e^{-px}. It was found that for $p > 10$, approximately, the numerical algorithm for the integration of stiff systems which we shall derive in this section has generally proved to be more efficient than the straightforward Taylor series algorithm with the ratio of the time taken by Taylor series to that taken by the algorithms derived later in this section becoming progressively larger as p increases. This "engineering" approach is necessarily imprecise due to the lack of precision in the definition of stiffness. We are not, of course, saying that a stiffness ratio of 10 is the precise point at which the straightforward Taylor series scheme should be abandoned but merely that the procedure just described has been found to be a convenient one in practice. If initially we find that a given system is non-stiff we may integrate forward using CTSS, which has been found to be efficient over a large class of non-stiff problems (Barton *et al.*, 1971b). We then need to re-examine our system for stiffness only if there is a substantial decrease in the size of the steplength of integration which is chosen automatically by the system. If on the other hand we have found a given system to be stiff the next problem is to determine which of the stiff components is the one causing the most trouble at a given point; i.e. to determine which component of the system has the smallest *time constant* associated with it at that particular point. The CTSS which uses Taylor series truncated after i terms, where i is an input parameter which may be chosen by the programmer, uses a slight modification of the standard device for choosing the steplength h whereby h is chosen to satisfy the inequality

$$h \le \left\{ \frac{(i!)\varepsilon}{y_M^i} \right\}^{1/i}.$$

Here y^i_M is the maximum in modulus of y^i_j, the ith derivative of y_j w.r.t. x for $j = 1, 2, \ldots, s$, and ε is the absolute accuracy required at each step. It follows immediately that for a given value of ε the stepsize of integration at any point x_c is determined by that component y_j which has the largest i^{th} derivative at x_c and it is this component which may be regarded as the one causing most trouble at x_c. It will generally be the case that if the jth component of (4.29) is the one causing the most trouble at x_c then the jth row of $\mathbf{A}(x_c)$ will contain some large elements with $a_{jj}(x_c)$, in particular, being large in modulus ($|a_{jj}(x_c)| > 10 + \sum_{i \neq j} |a_{ji}(x_c)|$ say). In this section we shall *only* be interested in systems of the form (4.29) for which $a_{jj}(x_c)$ is large in modulus if the jth component of (4.29) is the one causing most trouble at x_c since it is only systems of this form to which our algorithms, as they are at present formulated, may usefully be applied. We shall re-consider the implications of this important assumption at a later stage.

The situation now is that we have found our system to be stiff at the point x_c and it is the jth component which is causing the most trouble at this point. Writing the jth component out in full we have

$$y'_j(x_c) = \sum_{k \neq j}^{s} a_{jk}(x_c) y_k(x_c) + a_{jj}(x_c) y_j(x_c), \tag{4.30a}$$

where $\mathbf{y}(x_c) \equiv (y_1(x_c), y_2(x_c), \ldots, y_s(x_c))^T$. Suppose now that we have computed the approximate solution vector $\mathbf{y}(x_c)$ using the straightforward Taylor series method and that, as a result of the behaviour of the jth component, we decide to compute another approximation $\hat{y}_j(x_c)$ to the jth component of $\mathbf{y}(x_c)$ using an iteration scheme. In an attempt to eliminate the stiff component of (4.30a) we use the iteration scheme

$$y_j'^{(t-1)}(x_c) = \sum_{k \neq j}^{s} a_{jk}(x_c) y_k^{(t+1)}(x_c) + a_{jj}(x_c) y_j^{(t)}(x_c), \qquad y_j'^{(0)}(x_c) = 0, \tag{4.30b}$$

where $[y_1^{(t+1)}(x_c), y_2^{(t+1)}(x_c), \ldots, y_{j-1}^{(t+1)}(x_c), y_{j+1}^{(t+1)}(x_c), \ldots, y_s^{(t+1)}(x_c)] \equiv [y_1(x_c), y_2(x_c), \ldots, y_{j-1}(x_c), y_{j+1}(x_c), \ldots, y_s(x_c)]$ for all t. (Note that these components of the solution $\mathbf{y}(x_c)$ do not change—it is only the component $y_j(x_c)$ which is being changed by (4.30b).) The reason we write our iteration scheme in this particular form, i.e. incorporating three iteration levels, is that it allows us to derive a sequence of relations of the form

$$y_j^{(t)}(x_c) = \sum_{k \neq j}^{s} a_{jk}^{(t)}(x_c) y_k(x_c),$$

where the $a_{jk}^{(t)}(x_c)$ are defined uniquely in terms of the elements $\{a_{jk}(x_c)\}$. Putting $t = 1$ in (4.30b) we obtain

$$y_j^{(1)}(x_c) = -\frac{1}{a_{jj}(x_c)} \sum_{k \neq j}^{s} a_{jk}(x_c) y_k^{(2)}(x_c)$$

and so

$$a_{jk}^{(1)}(x_c) = -\frac{a_{jk}(x_c)}{a_{jj}(x_c)}.$$

Formally differentiating this relation with respect to x we obtain

$$\begin{aligned}
y_j'^{(1)}(x_c) &= \sum_{k \neq j}^{s} [a_{jk}^{(1)\prime}(x_c) y_k^{(2)}(x_c) \\
&\qquad + a_{jk}^{(1)}(x_c) y_k'^{(2)}(x_c)], \\
&= \sum_{k \neq j}^{s} a_{jk}^{(1)\prime}(x_c) y_k^{(2)}(x_c) \\
&\qquad + \sum_{k \neq j}^{s} a_{jk}^{(1)}(x_c) \sum_{i=1}^{s} a_{ki}(x_c) y_i^{(2)}(x_c), \\
&= \sum_{k \neq j}^{s} a_{jk}^{(1)\prime}(x_c) y_k^{(2)}(x_c) \\
&\qquad + \sum_{k \neq j}^{s} a_{jk}^{(1)}(x_c) \sum_{i \neq j}^{s} a_{ki}(x_c) y_i^{(2)}(x_c) \\
&\qquad + \sum_{k \neq j}^{s} a_{jk}^{(1)}(x_c) a_{kj}(x_c) y_j^{(2)}(x_c).
\end{aligned}$$

Substituting into (4.30b) for $t = 2$ we have

$$\begin{aligned}
&\sum_{k \neq j}^{s} [a_{jk}^{(1)\prime}(x_c) y_k^{(2)}(x_c) + a_{jk}^{(1)}(x_c) \sum_{i \neq j}^{s} a_{ki}(x_c) y_i^{(2)}(x_c)] \\
&\qquad + \sum_{k \neq j}^{s} a_{jk}^{(1)}(x_c) a_{kj}(x_c) y_j^{(2)}(x_c) \\
&\qquad = \sum_{k \neq j}^{s} a_{jk}(x_c) y_k^{(3)}(x_c) + a_{jj}(x_c) y_j^{(2)}(x_c).
\end{aligned}$$

So

$$y_j^{(2)}(x_c) = \sum_{k \neq j}^{s} a_{jk}^{(2)}(x_c) y_k(x_c)$$

since $y_k^{(2)} = y_k^{(3)} = y_k$ for $k \neq j$ where

$$a_{jk}^{(2)}(x_c) = \frac{1}{\left\{\sum_{k \neq j}^{s} a_{jk}^{(1)}(x_c)a_{kj}(x_c) - a_{jj}(x_c)\right\}}\left\{a_{jk}(x_c) - a_{jk}^{(1)\prime}(x_c) - \sum_{i \neq j}^{s} a_{ik}(x_c)a_{ji}^{(1)\prime}(x_c)\right\}.$$

For general t we have

$$y_j^{(t)}(x_c) = \sum_{k \neq j}^{s} a_{jk}^{(t)}(x_c)y_k(x_c),$$

where

$$a_{jk}^{(t)}(x_c) = \left\{a_{jk}(x_c) - a_{jk}^{(t-1)\prime}(x_c) - \sum_{i \neq j}^{s} a_{ik}(x_c)a_{ji}^{(t-1)}(x_c)\right\} \Big/ \left\{\sum_{k \neq j}^{s} a_{jk}^{(t-1)}(x_c)\, a_{kj}(x_c) - a_{jj}(x_c)\right\}$$

In common with most of the iteration schemes which we have so far considered, the convergence properties of scheme (4.30b) seem difficult to analyse. It is, however, clear that if convergence does occur then it does so to a solution of the original problem (4.29).

Assuming that the iteration scheme (4.30b) converges with increasing t it does so to a relation of the form

$$\hat{y}_j(x_c) = \sum_{k \neq j}^{s} \hat{a}_{jk}(x_c)y_k(x_c). \tag{4.31}$$

If we now replace the jth component of (4.29) by the relation (4.31) (note that this involves replacing a differential relation by an algebraic relation) and we replace all y_j appearing on the r.h.s. of (4.29) by (4.31) we obtain the modified first order system

$$\begin{aligned}\hat{\mathbf{y}}'(x) &= \hat{\mathbf{A}}(x)\hat{\mathbf{y}}(x),\\ \hat{y}_j(x) &= \sum_{k \neq j}^{s} \hat{a}_{jk}(x)\hat{y}_k(x)\end{aligned} \tag{4.32}$$

at $x = x_c$ where $\hat{y}(x) = (\hat{y}_1(x), \hat{y}_2(x), \ldots, \hat{y}_{j-1}(x), \hat{y}_{j+1}(x), \ldots, \hat{y}_s(x))^T$ and where the elements of $\hat{\mathbf{A}}(x)$ are uniquely defined in terms of those of $\mathbf{A}(x)$. (Note the dimension of $\hat{\mathbf{y}}$ is $s - 1$.) We now define a vector $\bar{\mathbf{y}}(x)$ as $\bar{\mathbf{y}}(x) = (\hat{y}_1(x), \hat{y}_2(x), \ldots, \hat{y}_j(x), \ldots, \hat{y}_s(x))^T$. Our procedure now is to integrate the modified system (4.32) with initial conditions $\hat{\mathbf{y}}(x_c) = (y_1(x_c), y_2(x_c), \ldots, y_{j-1}(x_c), y_{j+1}(x_c), \ldots, y_s(x_c))^T$ one more step-length h_1(this steplength will be chosen automatically by the CTSS) to give a solution $\hat{\mathbf{y}}(x)$ which may be evaluated for all $x_c \leq x \leq x_c + h_1$. Once the vector $\hat{\mathbf{y}}(x)$ is known we may compute $\hat{y}_j(x)$ from the second of relations (4.32) and hence the vector $\bar{\mathbf{y}}(x)$ may be determined. The original system (4.29) may also be integrated forward one more steplength h_2 (also chosen automatically by the CTSS and hopefully $h_1 > h_2$!) to give a solution $\mathbf{y}(x)$ which may be evaluated for all $x_c \leq x \leq x_c + h_2$. The two solution vectors corresponding to systems (4.29) and (4.32) may now be compared at the point $x_1 = \min(x_c + h_1, x_c + h_2)$. If these two solutions are such that

$$\left|\frac{y_i(x_1) - \bar{y}_i(x_1)}{y_i(x_1)}\right| > \varepsilon$$

for at least one value of $i \in [1, s]$, where we assume that $y_i(x_1) \neq 0$ and ε is the prescribed local error tolerance at each step, then exactly the same procedure is carried out starting from x_1. In particular a new modified system (4.32) is derived at the point x_1 and this is integrated forward with the initial conditions $\hat{\mathbf{y}}(x_1) = (y_1(x_1), y_2(x_1), \ldots, y_{j-1}(x_1), y_{j+1}(x_1), \ldots, y_s(x_1))^T$. This procedure may be continued until a steppoint $\bar{x}_m$ is reached for which

$$\left|\frac{y_i(\bar{x}_m) - \bar{y}_i(\bar{x}_m)}{y_i(\bar{x}_m)}\right| < \varepsilon$$

for all $i \in [1, s]$.

We may now draw the conclusion that since both systems have the same numerical solution, relative to the prescribed degree of precision, at the point $\bar{x}_m$, the transient solution which we are seeking to eliminate has decreased to negligible proportions at this point. As a result only the modified system (4.32) needs to be integrated forward from now on. In practice it is safer to integrate both systems forward one more steppoint and to abandon the original system only if the two solutions obtained again differ by less than the prescribed amount. Exactly the same procedure may now be applied to the modified system (4.32) starting from the point $\bar{x}_m$. If it is found that this modified system is still stiff at the point $\bar{x}_m$ a further modified system containing one less transient may be derived using the procedure just described. If it is found that the modified system is no longer stiff at the point $\bar{x}_m$, system (4.32) may be integrated efficiently using the straightforward Taylor series system. It is important to note that when integrating the unmodified system in the forward direction using Taylor series it is necessary to check at each step that it is still the jth component of this system which is the one causing the most trouble. If it is found at any stage that it is the nth and not the jth component which is causing most trouble (i.e. the jth and n^{th} components change their order of dominance) then a new modified system needs to be derived by applying our iteration scheme to the nth rather than the jth component of the original system. When this has been done the new modified system and the original system are then integrated forward in the usual way until either their solutions are "sufficiently close" or until a component other than the nth is found to be the one causing most trouble.

This whole procedure is based mainly on practical experience and has little theoretical backing. No doubt systems can be found for which our procedure fails to give the required solution. However, practical experience has indicated that this algorithm does have important applications and, as we shall see later, it is a natural extension of certain other algorithms which have been widely used in practice for the integration of a restricted class of problems. We shall return to this point in Section 4.9.

Our complete algorithm may now be summarised in a convenient step by step form in the following way:

(1) Set $k = 0$ and define $\tilde{x} \equiv x_0$, $\mathbf{C}_k(x) \equiv \mathbf{A}(x)$ and $\mathbf{y}_k(x) \equiv \mathbf{y}(x)$.
(2) Compute those eigenvalues of $\mathbf{C}_k(x)$ with the largest and smallest real parts at $x = \tilde{x}$. Assuming that it is known that the system is inherently stable (i.e. all the eigenvalues of $\mathbf{C}_k(x)$ have negative real parts) compute the stiffness ratio R at $\tilde{x}$. (This step is not needed if it is known from certain physical considerations that the system is stiff.) If $R > 10$ go to step 5.
(3) Integrate forward one more steplength using the straightforward CTSS.
(4) If the steplength chosen becomes substantially reduced at the point x_r, say, go to step 2 with $\tilde{x} = x_r$. Otherwise return to step 3.
(5) Integrate the system $\mathbf{y}_k'(x) = \mathbf{C}_k(x)\mathbf{y}_k(x)$ forward one steplength $h_{1,k}$ using the straightforward CTSS truncated after i terms starting from $\tilde{x}$. If the end of the range of integration has been reached, stop, otherwise find out which component may be regarded as the one causing the most trouble (the stiffest) at the point $\tilde{x}$. Suppose that it is the j_k^{th} component which is found to be the stiffest at this point:
(6) Apply the iteration scheme (4.30b) to the j_k^{th} component of the system $\mathbf{y}_k' = \mathbf{C}_k(x)\mathbf{y}_k$ to obtain a modified j_k^{th} component at the point $\tilde{x}$ on convergence. (If convergence does not occur at all the algorithm must be abandoned!) Denote the system obtained by replacing the j_k^{th} component of the system $\mathbf{y}_k'(x) = \mathbf{C}_k(x)\mathbf{y}_k(x)$ with the modified j_k^{th} component by

$\mathbf{y}'_{k+1} = \mathbf{C}_{k+1}(x)\mathbf{y}_{k+1}$. Note that $\mathbf{y}_{k+1}$ will have dimension $s - k - 1$ and as well as the $s - k - 1$ simultaneous first order differential equations $\mathbf{y}'_{k+1} = \mathbf{C}_{k+1}(x)\mathbf{y}_{k+1}$ there will be $k + 1$ algebraic relations for the components which have been found to be stiff.

(7) Integrate the system $\mathbf{y}'_{k+1} = \mathbf{C}_{k+1}(x)\mathbf{y}_{k+1}$ one steplength $h_{2,k}$ using CTSS to give a solution which may be evaluated at all points in the range $(\tilde{x}, \tilde{x} + h_{2,k})$. Compare the two solution vectors (i.e. the solution obtained by integrating the differential equations together with that obtained using the algebraic relations) at the point

$$x_p = \min(\tilde{x}_k + h_{1,k}, \tilde{x}_k + h_{2,k}).$$

If they are not sufficiently close in the sense described earlier set $\tilde{x} = \tilde{x} + x_p$ and go back to step 5. Otherwise:

(8) Set $k = k + 1$ and go to step 2 with $\tilde{x} = \tilde{x} + x_p$.

We now note some important practical properties of this algorithm. First we see that if the original system is not stiff our algorithm reduces to the straightforward Taylor series algorithm which has been shown to be efficient over a large class of non-stiff problems. Since the higher order derivatives associated with the Taylor series algorithm are generated automatically by the CTSS it is possible to integrate quite large systems of equations using this approach. It is clear, however, that the algorithm which we have proposed in this section can be used in conjunction with any explicit integration procedure which provides the required facilities, such as automatic step control, error estimation, etc., and is not peculiar to Taylor series. Secondly we note that an important degree of freedom which we have at our disposal is the choice of the number, i, of terms which are to be computed before the Taylor series (which is in general infinite) is truncated. Experiments so far performed have only been for small systems for which we have set $i = 12$ and this seems to be a convenient practical procedure. For larger systems it would be expected that a value of i rather less than this would be more efficient and it would be useful if we could establish an approximate relationship between system size and optimal value of i for a range of error tolerances (cf. Barton *et al.*, 1971a, b for non-stiff systems). Finally we note that iteration scheme (4.30b) is only applied to one component at a time. In general, if this iteration scheme converges reasonably rapidly, the computational effort required to iterate (4.30b) to convergence will be considerably less than that required to integrate forward one step using Taylor series.

Finally in this section we present a numerical example to illustrate our algorithm. Our particular example has been considered in detail in Cash (1974) but we give an abbreviated discussion of it here since it illustrates all of the important practical points associated with our algorithm.

Example 4.2

Consider the approximate numerical integration of the moderately stiff system

$$\mathbf{y}' = \mathbf{A}(x)\mathbf{y}, \qquad \mathbf{y}(0) = \mathbf{y}_0,$$

where

$$\mathbf{A}(x) = \begin{bmatrix} -0{\cdot}1 & -49{\cdot}9 & 0 \\ 0 & -50 & 0 \\ 0 & 70 & -120 \end{bmatrix}, \qquad \mathbf{y}_0 = \begin{bmatrix} 2 \\ 1 \\ 2 \end{bmatrix}.$$

This system is a particularly simple one but it does tend to make our algorithm rather more transparent than it would be if a more complicated system were to be considered. As a result of applying step 2 of our algorithm we find that the stiffness ratio of our system is 1200 and so it is necessary to proceed to step 5. As a result of applying step 5 it is found that it is the third component of our system which is causing the most trouble initially. Applying iteration scheme (4.30b) to the third component of our system we obtain

$$y_3^{(1)} = \tfrac{7}{12}\, y_2^{(2)}$$

and so

$$y_3'^{(1)} = \frac{7}{12} y_2'^{(2)} = -\frac{7 \times 50}{12} y_2^{(2)}.$$

It now follows that

$$y_3^{(2)} = \frac{7}{12} y_2^{(3)} + \frac{7 \times 50}{12 \times 120} y_2^{(3)} = \frac{133}{144} y_2^{(3)}.$$

(Note $y_2^{(2)}(0) = y_2^{(3)}(0) = 1.$)

Continuing this procedure we obtain the relation $y_3(x) = y_2(x)$ on convergence. This leads us to the modified system

$$\mathbf{y}' = \mathbf{c}_1(x)\mathbf{y}, \qquad \mathbf{y}(0) = \mathbf{d},$$

where

$$\mathbf{c}_1(x) = \begin{bmatrix} -0{\cdot}1 & -49{\cdot}9 \\ 0 & -50 \end{bmatrix}, \qquad \mathbf{d} = \begin{bmatrix} 2 \\ 1 \end{bmatrix},$$

together with the additional relation $y_3(x) = y_2(x)$. Our procedure now is to integrate the original and the modified systems forward one step x_p and to compare the solution $[y_1(x_p), y_2(x_p), y_3(x_p)]^T$ of the original system with the solution $[\hat{y}_1(x_p), \hat{y}_2(x_p), \hat{y}_2(x_p)]^T$ of the modified system (note that the second and third components of the modified system are equal). If these two solutions are not sufficiently close in the sense described earlier, the same procedure is carried out again with the initial conditions for the original system being $[y_1(x_p), y_2(x_p), y_3(x_p)]^T$ at $x = x_p$ and the initial conditions for the modified system being $[y_1(x_p), y_2(x_p)]^T$ at $x = x_p$. This procedure is continued until the corresponding solution vectors differ by less than the prescribed amount at some point x_c. We denote the solution obtained at the point x_c using the straightforward CTSS by $\mathbf{y}(x_c) = [y_1(x_c), y_2(x_c), y_3(x_c)]^T$. At the point x_c we abandon the original system and only integrate the modified system which is given by

$$\begin{aligned} y_1' &= -0{\cdot}1y_1 - 49{\cdot}9\,y_2, && y_1(x_c)\text{ given,} \\ y_2' &= -50y_2, && y_2(x_c)\text{ given,} \\ y_3(x) &= y_2(x). \end{aligned}$$

By applying our algorithm to the solution of this problem it is found that this system is still stiff and it is the second component which is causing the most trouble. Applying iteration scheme (4.30b) to this system we find that $y_2(x) \equiv 0$ on convergence. This leads us to the new modified system

$$\begin{aligned} y_1' &= -0{\cdot}1y_1, && y_1(x_c)\text{ given,} \\ y_2 &= 0, \\ y_3 &= 0. \end{aligned} \tag{4.33}$$

Note that we have used the fact that $y_2 = 0$ to modify the first equation as well. These two systems are integrated forward in the usual way until the solution vectors are sufficiently close. If this happens at some point $\bar{x}$ only the system

$$\begin{aligned} y_1' &= -0{\cdot}1y_1, && y_1(\bar{x})\text{ given,} \\ y_2 &= y_3 = 0 \end{aligned}$$

is integrated forward from the point $\bar{x}$. Some results comparing this algorithm with various other more standard ones are given in Cash (1974).

4.8 Extension to the non-linear case

In this section we consider the extension of some of the algorithms developed in the previous section to deal with the numerical integration of certain classes of non-linear differential equations of the form

$$\mathbf{y}' = \mathbf{A}(x, y), \qquad \mathbf{y}(x_0) = \mathbf{y}_0. \tag{4.34}$$

The basic idea behind the iterative solution of non-linear equations is the same as for the linear case in that it is determined first of all whether or not the given system is stiff and, if it is found to be stiff, the stiffest component is recognised by examining certain higher order derivatives and removed by use of a suitable iteration scheme. Rewriting system (4.34) in the form

$$\begin{aligned} y_1' &= A_1(x, y) \\ y_2' &= A_2(x, y) \\ &\vdots \\ y_i' &= A_i(x, y) \\ &\vdots \\ y_s' &= A_s(x, y) \end{aligned}$$

and assuming that it is the ith component which is found to be the stiffest at a given point x_c we may formally differentiate the ith component of our system with respect to x to obtain the relation

$$y_i'' = \frac{\partial}{\partial x}[A_i(x, y)] + \frac{\partial}{\partial y}[A_i(x, y)]\frac{\mathrm{d}y}{\mathrm{d}x}. \tag{4.35}$$

We note that if the explicit dependence on the independent variable appearing in (4.34) is removed by incorporating an additional dependent variable in the usual way, this differentiation requires the computation of one row of the Jacobian matrix at each steppoint. We may now attempt to generate the required slowly varying component of (4.35) using the iteration scheme

$$y_i'^{(t)} = \frac{\left\{ y_i''^{(t-1)} - \dfrac{\partial}{\partial x}[A_i(x, y^{(t)})] - \sum_j \dfrac{\partial}{\partial y_j}[A_i(x, y^{(')})] y_j'^{(t)} \right\}}{\dfrac{\partial}{\partial y_i}[A_i(x, y^{(t)})]}, \qquad j \neq i, \tag{4.36}$$

together with the usual initial approximation $y_i'^{(0)} = 0$. As is to be expected this algorithm for the non-linear problem is considerably more difficult to analyse than the corresponding algorithm which was derived for the linear case and at present no analytic results regarding the convergence of (4.36) are available. Nevertheless our algorithm has proved to be useful for the solution of a large class of non-linear problems and we illustrate its behaviour on a particular problem later on in this section. A slightly different approach, which is rather similar to that adopted in the linear case, is to try to remove the stiffest component via the use of an iteration scheme of the form

$$y_i'^{(t-1)} = A_i(x, y^{(t)}), \qquad y_i'^{(0)} = 0. \tag{4.37}$$

Instead of being required to solve a differential equation for $y_i^{(t)}$ we now need to solve a non-linear algebraic equation and in some cases this may be an easier task. We normally prefer the use of (4.36), however, since the Taylor series scheme is of an ideal form for use in conjunction with this iteration. When solving the differential equation for $y_i^{(p)}$, arising from an application of (4.36), using Taylor series it will be necessary to compute $y_i''^{(p)}$ and so this quantity is available for use when using scheme (4.36) with $t = p + 1$. As a result of this we may make our iteration scheme fully automatic if we replace the original i^{th} component of our differential equation by the system

$$y_i'^{(1)} = -\left\{\frac{\partial}{\partial x}[A_i(x, y^{(1)})] + \frac{\partial}{\partial y_j}[A_i(x, y^{(1)})]y_j'^{(1)}\right\} \bigg/ \frac{\partial}{\partial y_i}[A_i(x, y^{(1)})]$$

$$y_i'^{(2)} = -\left\{\frac{\partial}{\partial x}[A_i(x, y^{(2)})] + \frac{\partial}{\partial y_j}[A_i(x, y^{(2)})]y_j'^{(2)} - y_i''^{(1)}\right\} \bigg/ \frac{\partial}{\partial y_i}[A_i(x, y^{(2)})]$$

. . . .

. . . .

where the number, $q = q(x)$, of equations which need to be added depends on the accuracy required. If for example we require an absolute accuracy of ε at each step the criterion for choosing q would be

$$|y_i^{(q)} - y_i^{(q-1)}| < \varepsilon.$$

As a result of adopting this approach the dimension of our original system is increased but this is usually acceptable since, in general, we would prefer to integrate a large system of (ultimately) non-stiff equations using a standard explicit method rather than to integrate a smaller system of stiff equations using an implicit method. Also we usually find that as the integration proceeds rather less additional equations are required. This is due to the fact that certain of the extra equations which have been added may be abandoned at a later stage since we usually find, for the ith component for example, that

$$|y_i^{(q-1)} - y_i^{(q-2)}| < \varepsilon$$

after a while and so the equation for $y_i^{(q)}$ may be dropped. We also note that, unlike in the linear case, it is usually necessary to integrate each equation for the corresponding iterate $y_i^{(t)}$ and this involves a good deal of extra computation compared with the linear case. We give an example of the practical applications of this algorithm by considering the problem

$$\begin{aligned} y_1' &= -0{\cdot}04y_1 + 10^4 y_2 y_3 & y_1(0) &= 1 \\ y_2' &= 0{\cdot}04y_1 - 10^4 y_2 y_3 - 3 \times 10^7 y_2^2 & y_2(0) &= 0 \\ y_3' &= 3 \times 10^7 y_2^2 & y_3(0) &= 0 \end{aligned}$$

which has been considered in some detail in Cash (1974). By computing the stiffness ratio of our system at the point $x = 0$ it is found that the system is stiff initially and an examination of the higher derivatives of y_i w.r.t. x reveals that it is the second component which is initially causing the most trouble. The modified second component corresponding to (4.36) is

$$y_2'' = -(10^4 y_3 + 6 \times 10^7 y_2)y_2' - (0{\cdot}04)^2 y_1 + 400y_2y_3 - 3 \times 10^{11} y_2^3$$

and we seek to generate the slowly varying component of the general solution of this equation using the iteration scheme

$$\begin{aligned} y_2''^{(t-1)} = &-(10^4 y_3 + 6 \times 10^7 y_2^{(t)})y_2'^{(t)} - (0{\cdot}04)^2 y_1 \\ &+ 400y_2^{(t)}y_3 - 3 \times 10^{11} y_2^{(t)3}, \qquad y_2^{(0)\prime\prime} = 0. \end{aligned}$$

The whole procedure may be made fully automatic if this iteration scheme is presented to the Taylor series system as

$$\begin{aligned} y_2' &= -[0{\cdot}0016y_1 - 400y_2y_3 + 3 \times 10^{11} y_2^3]/(10000y_3 + 6 \times 10^7 y_2) \\ w_2' &= -[0{\cdot}0016y_1 - 400w_2y_3 + 3 \times 10^{11} w_2^3 + y_2'']/(10000y_3 + 6 \times 10^7 w_2) \\ z_2' &= -[0{\cdot}0016y_1 - 400z_2y_3 + 3 \times 10^{11} z_2^3 + w_2'']/(10000y_3 + 6 \times 10^7 z_2) \end{aligned}$$

. . . .

The results obtained for the solution of this problem are given in detail in Cash (1974) and the interested reader is referred to this paper.

4.9 Connections with existing algorithms

In this section we discuss some of the connections between the algorithms derived in the previous two sections and the widely used *pseudo-steady-state approximation* (PSSA) method which in turn has close connections with *singular perturbation* techniques. There is now a very extensive literature on singular perturbation methods and the present section is not in any way meant to be an exhaustive survey of the current literature but merely serves to introduce some existing algorithms which have connections with the sort of approach that we have been considering in this chapter so far. The usual way of introducing PSSA and singular perturbation methods is to consider the coupled system

$$\mathbf{x}' = \mathbf{f}(\mathbf{x}, \mathbf{y}, \varepsilon),$$

$$\varepsilon\mathbf{y}' = \mathbf{g}(\mathbf{x}, \mathbf{y}, \varepsilon),$$

where $\mathbf{x}$ is a non-stiff vector of dimension i, $\mathbf{y}$ is a stiff vector of dimension j with $i + j = s$ and ε is a small parameter. It is a fundamental characteristic associated with the numerical solution of stiff systems that, in direct contrast to the non-stiff case, there are two small parameters present, namely the stepsize h and the stiffness measure ε (the smaller the value of ε the more stiff the system is). In the steady state or slowly varying region of the range of integration, where the stiff components are negligible, we wish to have a situation where $h \gg \varepsilon$. Standard explicit methods of solution generally require that $h = k\varepsilon$, where k is normally quite small, everywhere in the range of integration and this is clearly an unacceptable situation. Since ε is a small parameter throughout the range of integration one possible approach would be to expand the required solution in a power series in ε. The mathematical theory behind the construction of these asymptotic series is now well known and may be found in Wasow (1965) for example. The PSSA method results from setting $\mathbf{y}' = 0$ initially. (This may be distinguished from setting $\varepsilon = 0$ which can cause a much more substantial error (Ray, 1969).) The first procedure which we mention, which has been based on the PSSA approach, is the one developed by McMillan (1968). McMillan observed the now well known fact that singular perturbation and PSSA methods are often inapplicable in the transient phase of the region of integration and so the first part of his algorithm consists of integrating forward until the transient phase has passed. In the steady state phase he then integrates the slow components using a standard explicit method while the fast components are obtained using an asymptotic expansion of the form

$$y(t, \varepsilon) = \sum_{i=0}^{q} y_i \varepsilon^i + 0(\varepsilon^{q+1}).$$

By using an appropriate matching procedure McMillan is able to find expressions for the y_i and hence the fast components may be computed at the required steppoints using a relatively large value of h.

Dahlquist (1969) has studied a slightly different approach for the numerical integration of a system of equations written in the form

$$\mathbf{x}' = -\mathbf{A}\mathbf{x} + \mathbf{g}(t, \mathbf{x}, \mathbf{y}),$$

$$\mathbf{y}' = \mathbf{f}(t, \mathbf{x}, \mathbf{y}),$$

where it is assumed that $\mathbf{x} \in \mathbb{R}^{s_x}$, $\mathbf{y} \in \mathbb{R}^{s_y}$, $\mathbf{A}$ is a piecewise constant s_x by s_x matrix and

$$\|\mathbf{A}^{-1}\| \left\{ \left\| \frac{\partial \mathbf{f}}{\partial \mathbf{y}} \right\| + \left\| \frac{\partial \mathbf{g}}{\partial \mathbf{x}} \right\| + 2\left(\left\| \frac{\partial \mathbf{g}}{\partial \mathbf{y}} \right\| \cdot \left\| \frac{\partial \mathbf{f}}{\partial \mathbf{x}} \right\| \right)^{\frac{1}{2}} \right\} < 1.$$

Dahlquist considers first of all the case where $\mathbf{g} \equiv \mathbf{g}(t)$ is a function of t only. In this case

$$\mathbf{p}(t) = -\sum_{q=0}^{N} (-\mathbf{A})^{q+1} \mathbf{g}^{(q)}(t)$$

gives a good approximation to the particular solution for the $\mathbf{x}$ component as N increases. An alternative way of obtaining this particular solution is to use the iteration scheme

$$\mathbf{p}_{i+1}(t) = \mathbf{A}^{-1}\left(\mathbf{g}(t) - \frac{d\mathbf{p}_i(t)}{dt} \right), \qquad \mathbf{p}_0(t) = 0,$$

which is similar to iteration scheme (4.37) considered earlier. One of the main differences between our approach and the approach developed by Dahlquist is that Dahlquist attempts to eliminate all of the stiff components simultaneously. This can only be done because he assumes that the original system of differential equations is in a certain particular form. In contrast, the algorithm developed in Section 4.7 seeks to eliminate one transient at a time. Although the algorithms which we have developed in Sections 4.7 and 4.8 do require the original system of differential equations to possess a special structure they do not go so far as to require the specially partitioned form considered by Dahlquist. Dahlquist also points out that his scheme is applicable in the case where $\mathbf{g} = \mathbf{g}(t, \mathbf{x}, \mathbf{y})$ and in this case there is a close relationship between Dahlquist's algorithm and the one proposed by McMillan. In common with the algorithm developed by McMillan, Dahlquist's algorithm requires different procedures to be used inside and outside the steady state phase. In the transient phase the system is integrated forward using a standard integration procedure. As soon as the steady state phase has been reached (and some numerical criterion must be used to determine when this is the case) a method called the SAPS method (smooth approximate particular solution) is used for the fast components, which are now contained in the vector **x**. This allows us to express our approximation to x in the form

$$\mathbf{x}_p(t) = \left(\mathbf{A}^{-1} - \mathbf{A}^{-2}\frac{\mathrm{d}}{\mathrm{d}t} + \mathbf{A}^{-3}\frac{\mathrm{d}^2}{\mathrm{d}t^2}\right)\mathbf{g}(t, \mathbf{x}_p, \mathbf{y}_p),$$

where, in the particular version which has been discussed by Dahlquist, the derivatives are approximated by corresponding finite difference terms. The equations corresponding to the slow components of the system may be integrated in an efficient manner using a standard finite difference method. If some form of local Richardson extrapolation is used a reasonable degree of accuracy may be obtained using this approach.

Another method of solution currently in use, which is closely related to the singular perturbation approach, is the steady state approximation method. To introduce this class of methods we again assume the original system to be in a partitioned form

$$\mathbf{x}' = \mathbf{f}(t, \mathbf{x}, \mathbf{y}),$$

$$\mathbf{y}' = \mathbf{g}(t, \mathbf{x}, \mathbf{y}),$$

where we note in particular that there is no explicit appearance of a stiffness parameter ε. We assume that the vector **x** corresponds to the relatively slowly varying components of the solution and **y** corresponds to those components which are rapidly varying originally but which soon reach a steady state. In the steady state phase of the region of integration the solution of our system can usually be closely approximated by the solution of the "steady state" system

$$\mathbf{x}' = \mathbf{f}(t, \mathbf{x}, \mathbf{y}).$$

$$0 = \mathbf{g}(t, \mathbf{x}, \mathbf{y}),$$

providing the way in which the original system is partitioned is suitably chosen. In order to illustrate this approach we consider the following stiff problem originally proposed by Robertson (1966)

$$
\begin{aligned}
y_1' &= -0{\cdot}04y_1 + 10^4 y_2 y_3, \\
y_2' &= 0{\cdot}04y_1 - 10^4 y_2 y_3 - 3 \times 10^7 y_2^2, \\
y_3' &= 3 \times 10^7 y_2^2,
\end{aligned}
\tag{4.38}
$$

with the initial conditions $y_1(0) = 1$, $y_2(0) = y_3(0) = 0$. This problem, which has been considered in detail in Cash (1974), has a Jacobian with eigenvalues $0, 0, -0{\cdot}04$ at $t = 0$ and as $t \to \infty$ these eigenvalues approach $0, 0, -10^4$. It was shown in Cash (1974) that it is the second component of this system which may be regarded as the one causing the most trouble initially. A steady state approximation to the solution of this problem would correspond to solving the system

$$
\begin{aligned}
y_1' &= -0{\cdot}04y_1 + 10^4 y_2 y_3, \\
0 &= 0{\cdot}04y_1 - 10^4 y_2 y_3 - 3 \times 10^7 y_2^2, \\
y_3' &= 3 \times 10^7 y_2^2.
\end{aligned}
\tag{4.39}
$$

The eigenvalues of this new system at $t = 0$ are given by $0, -0{\cdot}04$ and for large t they tend to $0, 0$. As a direct result of this, the problem of stiffness is eliminated and any standard explicit method with a sufficiently small value of h will be suitable for the efficient numerical integration of (4.39). This method of solution, if considered as a general purpose algorithm, does have some severe practical complexities (which the algorithms discussed in Sections 4.7, 4.8 attempt to overcome). Amongst the more serious of these are the following:

(1) We note first of all that the given initial conditions may not be consistent with the second of equations (4.39) being satisfied (this is the case with the example which we have considered) and so it may be necessary to modify the initial conditions in some way to take account of this. This problem will also arise if the variables are constrained to satisfy some sort of conservation law as is often the case in chemical kinetic problems.

(2) An obvious difficulty with this approach is that if we apply our steady state approximation for small values of t we will be attempting to approximate a non-steady state solution by a steady state approximation and this can lead to enormous errors.

(3) Another practical difficulty is that if we decide to integrate our problem in the forward direction using a standard integration method until the steady state phase has been reached, we have to derive a practical procedure which

allows us to recognise where the transient phase may be regarded as ending and the steady state phase begins.

(4) A final practical difficulty which we mention is one which is present with all the steady state or singular perturbation algorithms we have so far considered and it is that we need to assume our original problem is in a partitioned form so that we are able to identify the fast and the slow components of the solution.

The algorithm which was proposed in Section 4.7 and extended in Section 4.8 is able to overcome all of these problems, assuming of course that iteration scheme (4.30b) and the corresponding one for non-linear equations converge, as we now explain. Problem 4 is overcome by eliminating the stiff components one at a time and recognising the stiffest component at any point x_j as being that component with the largest derivative of a given order at x_j. We require our original system to have a particular structure to ensure that our iteration schemes converge but we do not require any *a priori* knowledge telling us which components of the system are stiff. This is a fundamental difference between the two approaches and is worth stating again explicitly. The singular perturbation and PSSA methods which we have so far considered require the original system to be in partitioned form so that the fast components can be distinguished from the slow components. However, the iterative algorithm derived in Section 4.7 requires the original system to have a special structure solely for the purpose of ensuring that the relevant iteration schemes converge. Problems 1, 2 and 3 are overcome by computing the solution using a straightforward explicit method as well as the iterative method and the original system is only abandoned if the two solutions differ by less than a prescribed amount in some appropriate sense. It is also of interest to note that the approach described in Section 4.8 may be regarded as a "corrected" PSSA method. An examination of the iterative scheme defined by (4.37) reveals that the first approximation (with $t = 1$) corresponds to the PSSA method. Scheme (4.37) does not stop at $t = 1$, however, but computes solutions for $t = 2, 3, \ldots$ until two successive values differ by less than a prescribed amount. This approach may be used in regions where $\mathbf{y}' \neq 0$ but reduces to the PSSA method in the steady state ($\mathbf{y}' = 0$).

Finally in this section we mention certain other attempts which have been made to overcome some of the problems just listed and which, if successful, may be used with our algorithm as well. The point is that our algorithm may be regarded as consisting of various "components" which have been developed to overcome the implementation problems just described. It is, however, possible to replace any of these components by more efficient ones as they are developed and in view of this we point out some promising approaches currently being developed which, if successful, could be incorporated into our algorithm to make it still more efficient. Firstly we note that most of our problems, apart from that of identifying the stiffest component of our system, could be overcome if we were able to choose the initial conditions in an appropriate manner, i.e. so that the initial conditions are correct for the required steady state approximation. In this case our iterative method of solution would be applicable over the whole range of integration and it would not be necessary to integrate the original system as well using a standard explicit method. The problem of determining the initial conditions required by these smooth solutions has been investigated by Dahlquist (1974) and it is hoped that his analysis, when fully developed, will prove useful for the determination of the appropriate initial conditions for our algorithm. Another method of approach which is aimed at overcoming some of these implementation problems has been considered by Lapidus *et al*., (1974). To explain their approach we assume that the system is written in partitioned form

$$\mathbf{x}' = \mathbf{f}(\mathbf{x}, \mathbf{y}),$$

$$\mathbf{y}' = \mathbf{w}(\mathbf{x}, \mathbf{y}),$$

where the vector $\mathbf{y}$ contains all of the stiff components. Lapidus *et al.* then identify this partitioned system with the system

$$\mathbf{x}' = \mathbf{f}(\mathbf{x}, \mathbf{y})$$

$$\varepsilon \mathbf{y}' = \mathbf{g}(\mathbf{x}, \mathbf{y})$$

by defining a local stiffness parameter ε which is such that (locally) $\mathbf{g} = \varepsilon \mathbf{w}$. Having done this they expand the solution in powers of ε up to order 1 in terms of what they call inner $(\mathbf{X}_0, \mathbf{X}_1, \mathbf{Y}_0, \mathbf{Y}_1)$ and outer $(\mathbf{x}_0, \mathbf{x}_1, \mathbf{y}_0, \mathbf{y}_1)$ variables. These expansions take the form

$$\mathbf{x} \sim \mathbf{X}_0 + \varepsilon \mathbf{X}_1 + \mathbf{x}_0 + \varepsilon \mathbf{x}_1,$$

$$\mathbf{y} \sim \mathbf{Y}_0 + \varepsilon \mathbf{Y}_1 + \mathbf{y}_0 + \varepsilon \mathbf{y}_1.$$

Explicit expressions for these inner and outer variables for general systems may be found by referring to the paper by Aitken and Lapidus (1974). Lapidus *et al.* note that, as usual, this method of solution is not valid in the transient phase and so some method suitable for determining when the transient phase has passed must be developed. The procedure which they propose for overcoming this problem is rather different from any which we have already considered and rather than produce it here we refer the interested reader to their original paper (Lapidus *et al.*, 1974).

Having given this brief survey we shall leave this problem but the problem of determining and extending the region of applicability of PSSA/singular perturbation/iteration methods is an interesting one which seems to merit further attention.

4.10 Linear second order O.D.E.s re-considered

To conclude this chapter we consider a small class of miscellaneous techniques which may be regarded as extensions of some of those developed in previous sections. We are mainly concerned with the numerical solution of linear second order differential equations and we start off by considering the boundary value problem

$$y'' + p(x)y' + q(x)y = 0, \qquad y(x_0) = y_0, \qquad y(x_N) = y_N, \tag{4.40}$$

which we assume to be stiff and which is such that $|p(x)|$ is large everywhere in the region of integration. The general solution, $y(x)$, of equation (4.40) may be written in the form

$$y(x) = c_1 y_1(x) + c_2 y_2(x), \tag{4.41}$$

where $y_1(x)$ and $y_2(x)$ are any two linearly independent solutions of (4.40) and the c_i are arbitrary constants. We assume that these basis solutions, $y_i(x)$, exist and may be chosen so that they are both approximately exponentials of the form $y_i(x) = y_0 e^{-\lambda_i x}$, $i = 1, 2$, with $|\lambda_1| \gg |\lambda_2|$. The relatively slowly varying component of (4.41), i.e. $y_2(x)$, may be computed using the iterative scheme (4.6) described in Section (4.1). In general, however, the required solution of (4.40) will contain a non-zero multiple of $y_1(x)$ and the problem still remains to generate this solution with an iterative scheme using a large steplength of integration, i.e. similar in size to that used to generate $y_2(x)$. To do this we note that at any point x in the range of integration we may compute λ_1, λ_2, which we assume are locally constant, as the solution of the quadratic equation

$$\lambda^2 + p(x)\lambda + q(x) = 0$$

which we assume has the distinct solutions $\lambda_1(x)$, $\lambda_2(x)$ which are real. If we now assume that the two basis solutions $y_1(x)$ and $y_2(x)$ of (4.40) do not change their order of dominance over the given range of integration, and in particular this will be the case if the solutions are approximately exponentials behaving like $y_0 e^{\lambda_i x}$, we may compute the rapidly varying solution $y_1(x)$ using scheme (4.6) by first making a transformation in the dependent variable $y(x)$. In order to determine a suitable transformation we first define a function $\lambda(x)$ by

$$\lambda(x) = max\,(|\lambda_1(x)|, |\lambda_2(x)|).$$

Since $\lambda(x)$ is defined at all steppoints in the range of integration we may define a constant λ as

$$\lambda = \max_x \lambda(x).$$

If we now make the transformation $z(x) = y(x)e^{\lambda x}$ we obtain a second order equation of the form

$$z''(x) + r(x)z'(x) + s(x)z(x) = 0, \tag{4.42}$$

which has two linearly independent solutions defined locally by the relations

$$z_1(x) = A_1 \exp\left(\int_0^x \lambda_1(\bar{x})\,d\bar{x} + \lambda x\right),$$

$$z_2(x) = A_2 \exp\left(\int_0^x \lambda_2(\bar{x})\,d\bar{x} + \lambda x\right).$$

We note that the constant λ has been specially chosen so that $z_1(x)$ is the least rapidly varying solution of (4.42). This solution $z_1(x)$ may now be computed from (4.42) using the iteration scheme

$$(z''/z)^{(t-1)} + r(x)(z'/z)^{(t)} + s(x) = 0, \qquad z''^{(0)} = 0. \tag{4.43}$$

Having obtained $z_1(x)$ we may obtain the required rapidly varying solution $y_1(x)$ using the reverse transformation $y_1(x) = z(x)e^{-\lambda x}$. Once the two solution sequences $y_1(x)$ and $y_2(x)$ are available we may determine the correct linear combination which satisfies the given boundary conditions and then we are able to determine the required numerical solution. We note that this procedure has been developed on the assumption that $\lambda(x)$ is a slowly varying function of x. If this is not the case it is often more efficient to divide the range of integration up into segments $s_j = [x_j, x_{j+1}]$ and to use a sequence λ_j which is such that

$$\lambda_j = \max_{x \in s_j} \lambda(x).$$

The adopted procedure would then be to make the transformation $z(x) = y(x)e^{\lambda_j x}$ in the segment s_j.

We now illustrate this approach by considering the particular problem

$$y'' + \left(\frac{400(x+1)^2 - 21}{20x + 19}\right)y' + \left(\frac{400x^2 + 780x + 360}{20x + 19}\right)y = 0$$

in the range $0 \leq x \leq 2$ with the boundary conditions $y(0) = 1 + e^{-10}$, $y(2) = e^{-2} + e^{-90}$. We compute the solution of this problem on an equi-spaced grid with grid spacing $h = 0{\cdot}1$. In order to find the slowly varying component, $y_2(x)$, of this problem we apply two iterations of scheme (4.6) and integrate the resulting first order ordinary differential equation using the backward Euler rule. The results obtained for the component $y_2(x)$ are given in Table 4.3. By computing the local eigenvalues of this system it is found that the function $\lambda(x)$ increases monotonically from $+18{\cdot}947368$ at $x = 0$ to $59{\cdot}661017$ at $x = 2$. The value of $\lambda(x)$ at all step points in the range of integration is given in Table 4.3.

Since $\lambda(x)$ is not a particularly slowly varying function of x, the range of integration is divided up into segments $s_j = [5(j-1), 5j-1], j = 1, 2, 3$, with $s_4 = [15, 21]$. By making the transformation $y(x) = z(x)\exp(-\lambda_j x)$ in the segment s_j, a differential equation for $z(x)$ is obtained and this is reduced to a linear first order ordinary differential equation by using two applications of iteration scheme (4.6). This first order equation is integrated using the backward Euler rule and on using the reverse transformation $z(x) = y(x)\exp(\lambda_j x)$ in the segment s_j a solution $y_1(x)$ is obtained and this solution is given in Table 4.3. By solving the two equations

$$Ay_1(0) + By_2(0) = 1 + e^{-10}$$

$$Ay_1(2) + By_2(2) = e^{-2} + e^{-90}$$

for A and B the required solution $y(x) = Ay_1(x) + By_2(x)$ is determined and the values of this solution together with the true solution are given in Table 4.3. As can be seen there is an error of about $0{\cdot}07$ initially when the term $e^{-10(x+1)^2}$ still has some significance, and so there is a large local truncation error in the backward Euler rule, but as x increases the solution obtained becomes increasingly accurate.

Finally in this chapter we consider the extension of some of our techniques to the solution of the special linear second order ordinary differential equation

$$y'' + f(x)y = g(x), \tag{4.44}$$

where the function $f(x)$ is large in modulus for all x in the range of integration. We are concerned only with the case where we wish to generate the particular

Table 4.3

x	$y_2(x)$	$\lambda(x)$	$y_1(x)$	$y(x)$	True solution
0·0	1·000000	−18·947368	$1{\cdot}000000 \times 10^{0}$	1·000045	1·000045
0·1	0·909716	−21·047619	$0{\cdot}134535 \times 10^{0}$	0·837305	0·904843
0·2	0·827464	−23·130435	$0{\cdot}147505 \times 10^{-1}$	0·751541	0·818731
0·3	0·752571	−25·200000	$0{\cdot}139868 \times 10^{-2}$	0·682396	0·740818
0·4	0·684405	−27·259259	$0{\cdot}132441 \times 10^{-3}$	0·620480	0·670320
0·5	0·622377	−29·310345	$0{\cdot}248769 \times 10^{-4}$	0·564237	0·606531
0·6	0·565945	−31·354839	$0{\cdot}123748 \times 10^{-5}$	0·513095	0·548812
0·7	0·514612	−33·393939	$0{\cdot}510838 \times 10^{-7}$	0·466537	0·496585
0·8	0·467921	−35·428571	$0{\cdot}181903 \times 10^{-8}$	0·424208	0·449329
0·9	0·425457	−37·459459	$0{\cdot}586266 \times 10^{-10}$	0·385710	0·406570
1·0	0·386838	−39·487179	$0{\cdot}184686 \times 10^{-11}$	0·350700	0·367879
1·1	0·351719	−41·512195	$0{\cdot}338254 \times 10^{-13}$	0·318862	0·332871
1·2	0·319784	−43·534884	$0{\cdot}516314 \times 10^{-15}$	0·289910	0·301194
1·3	0·290745	−45·555556	$0{\cdot}678894 \times 10^{-17}$	0·263584	0·272532
1·4	0·264341	−47·574468	$0{\cdot}794148 \times 10^{-19}$	0·239646	0·246597
1·5	0·240332	−49·591837	$0{\cdot}857532 \times 10^{-21}$	0·217880	0·223130
1·6	0·218502	−51·607843	$0{\cdot}577889 \times 10^{-23}$	0·198090	0·201897
1·7	0·198654	−53·622642	$0{\cdot}325019 \times 10^{-25}$	0·180095	0·182684
1·8	0·180607	−55·636364	$0{\cdot}157331 \times 10^{-27}$	0·163735	0·165299
1·9	0·164199	−57·649123	$0{\cdot}673309 \times 10^{-30}$	0·148860	0·149569
2·0	0·149281	−59·661017	$0{\cdot}261513 \times 10^{-32}$	0·135335	0·135335

solution of (4.44) and we assume that this particular solution is slowly varying compared with the unwanted complementary functions in some appropriate sense. An important practical application where this problem arises is in cases where these complementary functions are highly oscillatory and so, in a sense, the problem is poorly posed. We attempt to generate the required solution of (4.44) using the iteration scheme

$$y''^{(t-1)} + f(x)y^{(t)} = g(x), \tag{4.45}$$

where as usual we make the initial approximations $y''^{(0)} = 0$. Applying this scheme for increasing values of the integer t we obtain a sequence of approximations of the form

$$y^{(t)}(x) = A_t(x),$$

where

$$A_1(x) = g(x)/f(x)$$

and

$$A_t(x) = A_1(x) - \frac{1}{f(x)} A''_{t-1}(x) \tag{4.46}$$

This scheme may be continued until a maximum allowable number of iterates has been computed or until the sequence of iterates converges to the prescribed degree of precision ε in the sense that

$$|A_t(x) - A_{t-1}(x)| < \varepsilon$$

for all x in the range of integration. We again note that our algorithm is based on successive differentiation and so may be used in conjunction with an algebra system. As an example, which has also been considered as a test equation for oscillatory problems by Miranker and Wahba (1976), we consider the equation

$$y'' + \lambda^2 y = \lambda^2 \sin x. \tag{4.47}$$

where λ is a large constant. This single equation corresponds to the first order system of equations

$$\mathbf{z}' = \begin{pmatrix} 0 & 1 \\ -\lambda^2 & 0 \end{pmatrix} \mathbf{z} + \begin{pmatrix} 0 \\ \lambda^2 \sin x \end{pmatrix}. \tag{4.48}$$

The eigenvalues of the Jacobian matrix of this system are $\pm i\lambda$ and so when λ is large the complementary functions correspond to highly oscillatory solutions which are practically space filling. As a result of this the solution of (4.47) with given initial conditions which are such as to specify the slowly varying particular solution

$$y(x) = \frac{\sin x}{1 - \dfrac{1}{\lambda^2}}$$

is an ill-posed problem. We now consider the integration of this problem using scheme (4.46) where the differentiations are performed algebraically.

Example 4.3

Consider the solution of (4.47) using the iteration scheme (4.46). Then $y^{(1)} = \sin x$ and so $y''^{(1)} = -\sin x$.

Substituting into (4.45) for $t = 2$ we obtain

$$-\sin x + \lambda^2 y^{(2)} = \lambda^2 \sin x \quad \text{and so} \quad y^{(2)} = \sin x\left[1 + \frac{1}{\lambda^2}\right].$$

We now have $y''^{(2)} = -\sin x\left[1 + \dfrac{1}{\lambda^2}\right]$ and so

$$-\sin x\left[1 + \frac{1}{\lambda^2}\right] + \lambda^2 y^{(3)} = \lambda^2 \sin x$$

i.e.

$$y^{(3)} = \sin x\left[1 + \frac{1}{\lambda^2} + \frac{1}{\lambda^4}\right].$$

Continuing this procedure we obtain for general t

$$y^{(t)} = \sin x \sum_{i=0}^{t-1} \frac{1}{\lambda^{2i}}$$

and clearly $\lim_{t\to\infty} y^{(t)} = \dfrac{\sin x}{1 - \dfrac{1}{\lambda^2}}$ as required.

References

Aitken, R. C., and Lapidus, L., (1974). An effective numerical integration method for typical stiff systems. *A.I.Ch.E.*, **20**, 368–375.

Barton, D., Bourne, S., and Horton, J., (1970). The structure of the Cambridge Algebra System. *The Computer Journal*, **13**, 243–247.

Barton, D., Willers, I. M., and Zahar, R. V. M., (1971a). Taylor Series methods for ordinary differential equations—an evaluation. *In* "Mathematical Software" (J. R. Rice, Ed.). Academic Press, New York, pp. 369–390.

Barton, D., Willers, I. M., and Zahar, R. V. M., (1971b). The automatic solution of systems of ordinary differential equations by the method of Taylor Series. *The Computer Journal*, **14**, 243–248.

Cash, J. R., (1972). The numerical solution of linear stiff differential equations. *Proc. Cam. Phil. Soc.*, **71**, 505–515.

Cash, J. R., (1974). The numerical solution of systems of stiff ordinary differential equations. *Proc. Cam. Phil. Soc.*, **76**, 443–456.

Cash, J. R., (1975). The numerical solution of linear stiff boundary value problems. *The Computer Journal*, **18**, 371–372.

Cash, J. R., (1977a). A note on the iterative solution of recurrence relations. *Numer. Math.*, **27**, 165–170.

Cash, J. R., (1977b). A class of iterative algorithms for the integration of stiff systems of ordinary differential equations. *JIMA*, **19**, 325–335.

Cash, J. R., (1977c). On a class of cyclic methods for the numerical integration of stiff systems of O.D.E.s. *BIT*, **17**, 270–280.

Cash, J. R., (1978a). High order methods for the numerical integration of ordinary differential equations. *Numer. Math.*, **30**, 385–409.

Cash, J. R., (1978b). A note on the computational aspects of a class of implicit Runge–Kutta procedures. *JIMA*, **20**, 425–443.

Cash, J. R., (1978c). A note on a class of modified backward differentiation schemes. *JIMA*, **21**, 301–313.

Cash, J. R., (1978d). An extension of Olver's method for the numerical solution of linear recurrence relations. *Math. Comp.*, **32**, (142), 497–510.

Copson, E. T., (1970). On a generalisation of monotonic sequences. *Proc. Edinburgh Math. Soc.*, **17**, 159–164.

Dahlquist, G. G., (1963). A special stability problem for linear multistep methods. *BIT*, **3**, 27–43.

Dahlquist, G. G., (1969). A numerical method for some ordinary differential equations with large Lipschitz constants. *In* "Information Processing 68" (A. J. H. Morrell, Ed.). North Holland Publishing Company, Amsterdam, pp. 183–186.

Dahlquist, G. G., (1974). The sets of smooth solutions of differential and difference equations. *In* "Stiff Differential Systems" (R. A. Willoughby, Ed.). Plenum Press, New York, pp. 67–80.

Ehle, B. L., (1969). On Padé Approximations to the Exponential Function and *A*-Stable Methods for the Numerical Solution of Initial Value Problems. Res. Rep. No. CSRR 2010, University of Waterloo, Dept. Applied Analysis and Computer Science.

Fox, L., (1957). "The Numerical Solution of Two-Point Boundary Value Problems in O.D.E.s." Oxford University Press.

Gautschi, W., (1967). Computational Aspects of Three-term Recurrence Relations." SIAM Rev. 9, pp. 24–82.

Gear, C. W., (1969). The automatic integration of stiff ordinary differential equations. *In* "Information Processing 68" (A. J. H. Morrell, Ed.). North Holland Publishing Company, Amsterdam, pp. 187–193.

Henrici, P., (1962). "Discrete Variable Methods in Ordinary Differential Equations." John Wiley and Sons, New York.

Hull, T. E., (1974). The validation and comparison of programs for stiff systems. *In* "Stiff Differential Systems" (R. A. Willoughby, Ed.). Plenum Press, New York, pp. 151–164.

Hull, T. E., and Newbery, A. C. R., (1962). Corrector formulas for multistep integration methods. *SIAM*, **10**, 351–369.

Isaacson, E., and Keller, H. B., (1966). "Analysis of Numerical Methods." John Wiley and Sons, New York.

Lambert, J. D., (1973). "Computational Methods in Ordinary Differential Equations." John Wiley and Sons, New York.

Lambert, J. D., (1974). Two unconventional classes of methods for stiff systems. *In* "Stiff Differential Systems" (R. A. Willoughby. Ed.). Plenum Press. New York. pp. 171–186.

Lapidus, L., Aitken, R. C., and Liu, Y. A., (1974). The occurrence and numerical solution of physical and chemical systems having widely separated time constants. *In* "Stiff Differential Systems" (R. A. Willoughby, Ed.). Plenum Press, New York, pp. 187–200.

Liniger, W., and Willoughby, R. A., (1967). "Efficient Numerical Integration Methods for Stiff Systems of Differential Equations." IBM Research Report RC–1970.

Loeb, A. M., and Schiesser, W. E. (1974). Stiffness and accuracy in the method of lines integration of partial differential equations. *In* "Stiff Differential Systems" (R. A. Willoughby, Ed.). Plenum Press, New York, pp. 229–243.

McMillan, D. B., (1968). Asymptotic methods for systems of differential equations in which some variables have short response times. *SIAM J. Appl. Math.*, **16**, 704–722.

Miller, J. C. P., (1952). "British Association for the Advancement of Science: Bessel Functions Part II, Mathematical Tables, Vol 10." Cambridge University Press.

Miller, J. C. P., (1966). Chapter 4 *In* "Numerical Analysis—An Introduction" (J. Walsh, Ed.). Academic Press, New York.

Miller, K. S., (1968). "Linear Difference Equations." W. A. Benjamin, New York.

Miranker, W. L., and Wahba, G., (1976). An averaging method for the stiff highly oscillatory problem. *Math. Comp.*, **30**, 383–399.

Nordsieck, A., (1962). On numerical integration of ordinary differential equations. *Math. Comp.*, **16**, 22–49.

Nordsieck, A., (1963). Automatic numerical integration of ordinary differential equations. *AMS Proc. Symp. Appl. Math.*, **15**, 241–250.

Oliver, J., (1968a). The numerical solution of linear recurrence relations. *Numer. Math.* **11**, 349–360.

Oliver, J., (1968b). An extension of Olver's error estimation technique for linear recurrence relations. *Numer. Math.*, **12**, 459–467.

Olver, F. W. J., (1967). Numerical solution of second order linear difference relations. *J. Res. NBS*, **71B**, 111–129.

Olver, F. W. J., and Sookne, D. J., (1972). Note on backward recurrence algorithms. *Math Comp.*, **26**, 941–947.

Ortega, J. M., and Rheinboldt, W. C., (1970). "Iterative Solution of Nonlinear Equations in Several Variables." Academic Press, New York.

Pereyra, V., (1967). Iterated deferred corrections for nonlinear operator equations. *Numer. Math.*, **10**, 316–323.

Ray, W. H., (1969). The quasi-steady-state approximation in continuous stirred tank reactors. *Canadian J. Chem. Eng.*, **47**, 503.

Richtmeyer, R. D., and Morton, K. W., (1967). "Difference Methods for Initial Value Problems." John Wiley and Sons, New York.

Robertson, H. H., (1966). The solution of a set of reaction rate equations. *In* "Numerical Analysis—An Introduction" (J. Walsh, Ed.). Academic Press, New York, pp. 178–182.

Shampine, L. F., and Gordon, M. K., (1975). "Computer Solution of Ordinary Differential Equations." W. H. Freeman and Co., San Francisco.

Stetter, H. J., (1973). "Analysis of Discretization Methods for Ordinary Differential Equations." Springer-Verlag, Berlin.

Urabe, M., (1970). An implicit one step method of high order accuracy for the numerical integration of ordinary differential equations. *Numer. Math.*, **15**, 151–164.

Wasow, W. (1965). "Asymptotic Expansions for Ordinary Differential Equations." Interscience, New York.

Watson, G. N., (1952). "Treatise on the Theory of Bessel Functions." Cambridge University Press.

Widlund, O. B., (1967). A note on unconditionally stable linear multistep methods. *BIT*, **7**, 65–70.

Willoughby, R. A., (Ed.). (1974). "Stiff Differential Systems." Plenum Press, New York.

Zahar, R. V. M., (1977). A mathematical analysis of Miller's algorithm. *Numer. Math.*, **27**, 427–447.

Index

Bold numbers indicate pages on which major references appear or where definitions are given.